FARMING TIMES

Farming Times

PAUL HEINEY

First published in Great Britain 1992
by H. F. & G. Witherby Ltd

Published in Large Print 1993 by Isis Publishing Ltd,
55 St. Thomas' Street, Oxford OX1 1JG,
by arrangement with Victor Gollancz Ltd.

British Library Cataloguing in Publication Data
Heiney, Paul
Farming Times. — New ed
I. Title
630.92

ISBN 1-85695-041-7

Printed and bound by Hartnolls Ltd, Bodmin, Cornwall
Cover designed by CGS Studios Ltd, Cheltenham

CONTENTS

I dedicate this book
to my wife Libby
without whom I would never
have had a farm of my own

(Dedicating a book to one's wife is, I am told, a typical thing for a farmer to do. Not only does it bring a soothing glow to the household, it costs nothing.)

ACKNOWLEDGEMENTS

This book is a collection of articles written for *The Times* and I am grateful, and flattered, that they should consider them worthy of publication. In particular, I would like to thank Andrew Harvey who first commissioned them; and Brigid Callaghan and Rose Wild who later took responsibility for them.

At home I have had an unfailing support team and in particular would like to thank Richard White, and his son Robert, who have cheerfully answered calls for help; and Guy Seinet whose knowledge and skill concerning sheep has saved me and my ewes much heartache.

Derek Filby, a true horseman, has also been a great support. He worked horses on the land after leaving school and gave up when the farmer insisted it was time for tractors to take over. Nearly thirty years later he arrived on our farm and gladly passed on the skills he had acquired in his farming days. Without him, our furrows would undoubtedly be less straight and our rows of corn even more wavy.

I must also thank Dilly Sharp who appears later in this book. He has not only provided me with plenty of his farming wisdom, but kept us all cheerful through some of the darkest farming days with his accounts of even worse disasters.

INTRODUCTION

For the first thirty-eight years of my life I had not the slightest interest in farming. For anyone to suggest that I might become a farmer would have been as ridiculous as tipping me to win the Grand National. Not only was I ignorant of farmers and what they did, I probably shared the common suspicion that modern farming was more about fiddling with figures than about the art of growing.

But my conversion was sudden and it happened by chance. I was invited to the annual dinner of the Suffolk Horse Society and sat between Roger and Cheryl Clark, both expert farriers and distinguished owners of Suffolk Punches. During the evening odd snippets of conversation revealed what it was that set this couple apart: they were also farmers. Unusual farmers these days: they cultivate over a hundred acres using carthorses rather than tractors.

At this stage I was merely curious, but after my first visit to their farm, deep in Constable country, I was drawn into their world. It is a heady atmosphere, which blends the might of working horses with the honest and skilful toil of the traditional farm labourer, but it was not only the romantic aspects that appealed to me. I sensed that here was an agriculture that was within the grasp of the common man; something on a more human scale than modern high-tech agriculture. It was also, by its pace and techniques, kinder to the land and the livestock. It treated the soil with a respect and sensitivity that all observation suggested were becoming increasingly rare in modern farming.

Watching the shimmering chestnut horses hauling the plough down ruler-straight furrows, smelling the richness of the newly turned earth and hearing the crowing of the seagulls as they swooped on to the furrows to steal worms, I began to long to be a farmer. This kind of farmer. I worked part-time with the Clarks for a year and wrote of my experiences (*Pulling Punches*, Methuen 1988) and as that year progressed it became clear that I would not be able easily to turn my back on the job.

The following year, with my wife and children, I circumnavigated Britain in our small yacht. On the lengthier parts of the voyage I found my mind wandering back to the fields and the meadows; and on our arrival home in the autumn of 1988 I broke the news to my wife that I wanted to become a farmer. Poor woman. When we first met I was living a bachelor life in a smart little mews flat in Kensington; when we married we lived in bijou Georgian houses in London. She did not complain when I eventually suggested we move to Suffolk in the early eighties, long before it became the fashionable thing to do, but I was far from certain how she would react to this latest wild suggestion.

Typically, she had seen it coming. Perhaps it was the way that on leaving the Clarks I told her I had bought a Suffolk Punch, and that when the horsebox arrived there were two of them in it. Perhaps it was the way I kept coming back from the farm auctions with harrows and muck-carts and tangled, rusty, precious old pieces of horse-drawn machinery. Maybe it was the cast-iron pig troughs or the turnip-slicing machine hidden behind the garden shed which gave her a clue. She bowed to the inevitable.

Of course, taking up farming meant a house move, which coincided with the slump in the housing market. This led to many anxious and frustrating months. The biggest hurdle of all was finding the farm of the right size, with the sort of house we could all live in, and the traditional buildings that we needed for our kind of farming. Most farm buildings now are great empty tin-roofed cathedrals, utterly unfit for tethering horses or sheltering small quantities of mixed animals. The old walled farmyards, too poky for big machines and intensive rearing, are turned into holiday cottage complexes, or merely fall into ruin. But after a struggle that lasted for nearly two years, to our great surprise we suddenly found our ideal farm. We arrived in the spring of 1990, financially overstretched, nervous and excited at the thought of what lay ahead. With more bravado than I felt inside, I wrote a piece for *The Times* about my intentions and beliefs. It attracted a volley of furious criticism from the conventional farming lobby, but an even larger volume of kindly advice and encouragement from those who, like me, saw merit in the old ways and thought that in ditching them, the farming community had thrown out the baby with the bathwater.

The Times asked for these weekly progress reports. At times — when stock escaped, cows died, recession undermined the country, the work overwhelmed me or the turnips came up all spotty — I have wondered whether we would last another six months, but as I write the farm is still in business.

Middleton, February 1992

CHAPTER ONE

Beginnings

Dreams, Machines and Moonshine

On my new farm there is a tumbledown barn that leans precariously away from the wind and creaks at every gust. There are rusty iron gutters along the edge of gaping roofs. Inside the crumbling buildings, wooden mangers are rubbed smooth where greedy bovine tongues once licked every ounce of corn from them. And in the soft red brick are scratched the initials of the men who, a century ago, did what I am going to try to do today. I am going to be a farmer.

Not a modern farmer. I am not going in for the current state-of-the-art agriculture which demands maximum return at whatever expense to land, animal or planet. I am not even going to be a reasonable mechanised organic farmer. I want to have the sort of farm of which children's books were written, where chickens scratched at the foot of haystacks, pigs rooted in corners of fields (for whatever it is that pigs root so earnestly for) and where lambs frolicked in meadows.

Stinking tractors won't get a look-in either: I have bought three mighty carthorses, Suffolk Punches, which will reap and sow, plough and mow, carthorses doing what they were bred for. And there I shall be, at the centre of it all, leaning over the five-bar gate dispensing dubious rural wisdom to passers-by. Think of Old MacDonald and his farm, and you've got it.

But my farm will be no joke. I know that many modern farmers will already be laughing their socks off at the thought that a smug ex-townie with his old horse can teach them anything about growing food. All I would say is that within the walls of my crumbling farmyard buildings lies a fossilised wisdom which is about to have its resurrection.

The farming clock has to be turned back to the days when farming made sense. Few people these days seem to like farmers very much, which is unhealthy. They feed us cheaply and plentifully and we ought to be able to trust them. But even during the 1989 "Food and Farming Year" in which they were supposed to boast of their achievements, all they could manage was to fuel suspicion as to their dubious practice of feeding live animals with dead ones, some of which were insane. Fifty years ago, farmers were perceived as jolly chaps who ploughed the fields and scattered till all was safely gathered in. No one sings hymns of praise about farming these days. It's a dark subject.

I think I can throw some light on it. Not as an agriculturalist or bio-scientist but as one of a growing band of people who think it is time the grain train hit the buffers. It has been running out of control since the Second World War when the need to produce food was desperate. Grassland, meadows and heath went under as Churchill ordered the ploughs onwards. But farmers never got out of the habit; the grain train ran away with them. The agro-chemical industry spurred them on to produce more and more without a thought for the consequences to the land or the food. It also made some of them rich. Now we have had that system we are just beginning to see the damage it has done. There now exist question marks over nitrates in drinking water, salmonella in eggs, mad-cow disease in beef, hormones in milk and antibiotics in bacon. Too many questions, I think, to which there is only one answer.

The men who scratched their initials in the soft brick of my tumbledown farm, "A.J.P. 1867" or "P. ELY 1892", were farming under a system which was better than any invented since, and from which we could still learn a lot. They worked in harmony with natural systems rather than trying to fight them. The muck their animals produced, to take one example, was not the embarrassment it is today. If I had not seen it with my own eyes, I would not have believed that slurry produced on farms in Holland (the consequence of highly intensive production of pig-meat) is regularly loaded into barges and taken for rides up and down the river Rhine because they cannot think of anything else to do with it. As far as I know, British farmers may well be taking theirs for rides round and round the M25.

"A.J.P.", whoever he was, would have laughed. He knew how to keep stock and feed his land for free. The simplicity was the beauty of the system. He grew the corn and saved the straw. When winter drew on, the stock were brought from the meadows into the farmyard for shelter and to be fed. Down their throats went some of his precious corn and, obligingly, the animal deposited the digested remains in neat little pats on the very straw that had carried the corn the summer long.

After six months of being trodden by ever heavier hooves in the farmyard, a miraculous transformation would have taken place; for when “A.J.P.” stuck his fork into the muck he found it had turned into dark, rich, nutritious, rotted compost. It had cost him nothing at all.

The horse and cart were sent for, and an army of men, and forkful by sticky forkful the precious matter was carted to the field to be ploughed in, to help grow more corn, to feed more stock, to produce more straw, to fill the yards in the winter. “E-i-e-i-o,” as Old MacDonald would have said. And so revolved a highly efficient and natural cycle. It was organic farming before anyone had come up with the idea.

Will it work today, on my farm? To the satisfaction of my critics, who will be many, I have to admit that it may not. Not on the small scale I can afford to do it. It is not because the natural system was in any way at fault: it is just that the figures don’t add up any more. Yields of corn on the old farms were much lower than today, labour was cheaper, there was plenty of it and farmers did not feel they had a right to a rich living. But we now pay farmers *not* to farm their land (it is called “set-aside”). Meanwhile we have Development Commissions which scratch their heads trying to create jobs in rural areas. We are paying huge sums to solve problems of our own creation. So, to prove that the old farming was the best farming for the land and the countryside, I am personally putting on my boots and taking to the soil. It is the principle that matters, and the pounds will have to take care of themselves. It will cost all my family’s spare income at first and cause some anxiety. But I don’t care.

The farm runs to only thirty-six acres. Not big, hardly room enough for an out-of-town Tesco, but it will do to prove the point. Thankfully it has been in good hands, for the fields still have their hedges, the trees are still standing and the first-class barn has not been converted into a second-class house.

In the valley in which my farm shelters were once some of the most lush and tasty grazing marshes in the area but the "advance" of farming took the cattle off them and put them in intensive units. Where once grew five-star fodder, invasive and unpalatable sedge has taken charge. We are joining forces with the Suffolk Wildlife Trust to restore them to verdant pasture: this will add a rented twenty-five acres to what we own, which will be useful.

By Suffolk standards, the countryside is hilly. From the highest point of the farm I can stand and smell, if not quite see, the sea. From the highest point, I hope, in a few years' time, to look down on a farm as it would have looked a hundred years ago: fields of corn grown organically without the help of synthetic fertilisers and pesticides, pastures rich in herbs and grasses that give the animals that graze them a glow of good health. Horses will be at work, too. In winter, Suffolk Punches might be carting hay or turnips to sheep. If it is June they will pull the mower that cuts the grass to make hay. We shall cart the hay to the farmyard and with our pitchforks build haystacks. Remember those? In the winter the horses will labour before the plough turning the used earth and making it new again.

I am not one of the soil's obvious sons. I began with gardening and leeks like telegraph poles springing from the organic garden soil fed with stolen horse-muck. We had onions like footballs, marrows of vulgar dimensions and occasional set-backs when caterpillars turned cabbage leaves into lace curtains and worms carved tunnels of love through otherwise promising potatoes.

Then fate took me on a casual visit to a small farm in Suffolk which, even at the end of the twentieth century, is surprisingly still worked by horse. I was hooked. I worked one day a week for a year as a farmhand, wrote a book about it, and grew to understand not only the way farming used to work, but more importantly, how the carthorse fitted into the grand scheme. It was a formative year, at times exhausting and dispiriting. You try carting heavy mangelwurzels on frosty, foggy mornings. But when we went to plough, it was like making music. Ploughing with horses is like no other experience that the land can offer. When it is going well and the horses, the man and the plough are in tune it is symphonic: the plough cuts straight and deep and the soil sings as it glides across the ploughbreast. The jangle of the harness underscores the whole piece and the heavy plod of the horses' hooves give it a rhythm. At the end of the furrow, you turn to see the landscape which you have scarred and then guide your horses round for the beginning of the next movement.

By some stroke of great good luck I had arrived at a point where my passions for the countryside, for working with carthorses and for growing food could fuse together. I decided to buy a farm.

In the meantime, I had considerable gaps in my knowledge to fill. Modern farming textbooks were clearly of no use so my education was gleaned from such diverse volumes as *Stephen's Book of the Farm: Second Edition,* 1877 and *Mayhew's Illustrated Horse Management,* 1890. I gather from the style supplements of the Sunday newspapers that we are judged by the books that we are seen to be reading in public; Martin Amis and Julian Barnes are pretty high on the list of desirables. Alas, there was no mention of *Humus and the Farmer —A Critical View* or *The Pioneering Pig*, two volumes from which I have recently found great inspiration.

I fell in the style index. Baffled media people began to get scripts returned with wisps of straw between the pages and with pig carcass prices scribbled up the margin. The envelopes smelt horribly because the foolish boy who brought the pig-feed put it down on the pile of letters. If, by the way, there is a publisher who finds a set of rubber rings for castrating sheep mixed in with a manuscript, they're quite expensive and I'd like them back.

But slowly I am learning about farming. And about farmers. I went to see one of the old timers the other day to try to buy off him a horse-drawn plough. He took me into the yard where it stood, and told me the full story. "It was my father's," he pined. "Good old plough it is, he taught me everything he knew behind that old plough. No, that's an old family friend really. I couldn't sell that . . ." and he took a long pause, and with a knowing look said, ". . . unless that was for a hundred pounds!" Sly old devil.

I shall need a touch of cunning like that if the money is not to run out entirely. I don't suppose I'll get much support from serious farmers. They will dismiss it all as fanciful romanticism, unaware that the sands of time are running out for the system they uphold. My extreme reaction is probably not the answer either, but it is a gesture.

If I have anyone on my side, I hope it will be the spirits of "A.J.P." and "P. ELY" for they, and men like them, created a science of farming that deserves better than to be dismissed as inefficient and sentimental rubbish. I am about to start on what for me will be a great adventure. As I put the collars on the horses and we take to the fields to work, I shall think of those men. I hope they will be with me on every long trudge down the furrow.

On the first night in our new farm, war broke out: a wicked skirmish that shook the old stable buildings to their dubious foundations.

On an ordinary farm, you can reasonably assume that if you buy a new tractor, it will stand quite happily with the old ones. The same is not true of carthorses.

Our new young horse, named Blue, had arrived in the afternoon and taken his place in the stable with our two old-stagers, Punch and Star. Nothing much happened. After feeding, and with dusk falling, I turned all three into our enclosed horse-yard for the night. That was their cue. There was a mighty roar from the normally docile Star and a buck and a kick from the usually idle Punch. Blue cowered in a corner. I put a heavy gate between me and them as they charged and galloped, kicked and bit.

Shaken, I retired to the house to find hostilities in the attic. The people who lived here before have only moved across the lane and their poor cat is confused: it thinks it still lives here. Our cats, on the other hand, seem to understand the laws of property only too well and were serving a violent eviction order. As I lay in my bed that first night there were feline fisticuffs to the right of me, equine anger to the left: stereophonic warfare.

The next morning dawned clear, sunny and bright. I

nervously went out to the yard, expecting bloody limbs and torn horseflesh and was met by the sight of three horses gently munching their hay like children who could not remember why they fell out in the first place. Relieved, I took a stroll up the farm.

I had always imagined that walking one's own farm would be one of life's great pleasures, but it is not. Like a true artist who only ever sees the faults in his work, I suspect a good farmer sees the weeds before the crop. I didn't have far to look. Pernicious nettles, thistles and docks all sprouted vigorously in the unseasonable early spring sunshine. I cursed the warm weather, and then silenced myself for sounding too much like a farmer. However, something will have to be done about weeds: impure inorganic thoughts flashed through my mind. But no, this is to be an organic farm and each and every weed will require painstaking personal attention.

I walked a little further and the ground beneath my feet changed. It was like walking off the beach and on to a motorway hard shoulder. A heavy tractor had clearly driven repeatedly over the same patch of ground and had compressed a reasonable bit of soil into a solid, unworkable lump of land; unworkable, that is, by horse.

On such a fine morning, my neighbours were already hard at work. In the distance a tractor and sprayer were zigzagging up and down a field and the taste of chemical borne on an unfortunate breeze was soon on my lips. I dived behind a hedge and thought. His right to spray is incontrovertible but so is mine to farm without chemicals. There will have to be some diplomacy across the hedge.

The air of tension that had hung over the farm since the stableyard fight was broken by the arrival of an encouraging letter from a friend. He quoted the warning to the farmers in Britten's *Paul Bunyan* (libretto, W. H. Auden):

> If there isn't a flood, there's a drought.
> If there isn't a frost, there's a heatwave.
> If it isn't the insects, it's the banks.
> You'll howl more than you'll sing.
> You'll frown more than you'll smile.
> You'll cry more than you'll laugh.
> But some people seem to like it.
> Let's get going.

Feeling better, I went to harness the horses for the first day's work on my own farm.

There is a gradual reawakening taking place on the farm, and it is nothing to do with the spring. I see it on the edge of the old mangers: dusty and dull when we first came, but now licked smooth and shiny by horses' lashing tongues. Carthorses seem to know when work is coming; they will stand idly in their stalls until they hear the rattle of the approaching harness chains, then start to hunt for the final fortifying grain of food. It is the horse's equivalent of one for the road.

Around the fields our cart-tracks are already changing. Tractors leave two deep tyre-ruts behind them with a mound between, but the repeated plod of a heavily shod carthorse wears away the ground down the middle of the track, leaving a shallow gully. Much easier for walking, and better for getting rid of rainwater, too. We shall appreciate it in the winter when muck has to be carted from the farmyard to the fields.

Gates and barn doors that creaked with age have responded to regular exercise and groan no more; gutters, freed of thirty years' worth of rotting autumn leaves, now chuckle to themselves when the rain comes.

If you think of our farm as a rusty old machine, I feel we have at least given it the first drop of oil.

But something is not quite right, and it has taken me many miles of furrow-walking to realise exactly what it is. We do not have any gulls following the plough.

Now, gulls are to newly turned furrows what young girls are to pop groups: they scream. But no gulls follow me: it can't be just the mild weather.

The answer lies in the soil. Gulls do not follow the plough out of some desire to live up to a chocolate-box representation of the countryside. They do it for food. They swoop down, squabble and pluck fat, succulent worms whose world has been turned upside down by the plough. If you have no worms, you get no gulls. No worms means dead soil, so gulls and living soil go together.

Some of our soil is very dead, and you can tell it from the colour. On smaller fields, old pastures where animals grazed and dunged for years, the soil is richer and blacker and as nourishing as Christmas pudding. But where the earth has been beaten into submission by the combined assault of heavy machinery and chemical feeding, it just sits there disabled, waiting to be fed.

As I understand it, intensive agriculture feeds the growing plant by applying nourishment in chemical form; organic farming feeds the soil beneath, by applying farmyard manure or compost. The plant then feeds from the enriched soil, as naturally as a baby at its mother's breast.

A lot of modern farmers know that what they are doing to the soil is wrong, and resent having to do it: a neighbour told me that every few years he needs several more horsepower from his tractor to pull the same plough through the same fields. The soil is dying, giving up the struggle. He knows it, but he has to pay the bills: the economics of modern farming do not leave much room for charity.

My present act of charity begins in a field where we are ploughing with a view to planting clover. Clover is a crop whose ability to fix invigorating nitrogen in the soil has made it the darling of the organic farming movement. If you follow a field of clover with a crop of corn, the corn won't need any fertiliser. Or so the theory goes.

And there is further value in it: clover becomes sweet hay, best made by the gentle, unhurried movement of horse-drawn farm machinery. Modern haymaking gear grabs it, throws it, and shakes off the leaf where much of the goodness lies. In the gentle caress of our slow and solid pre-war machinery, leaf and stalk make their way to the haystack together.

And when the hay is taken, the clover is ploughed under to release its natural fertiliser, rotting and encouraging the precious worms to turn and aerate the soil.

If we have ploughed well and given the seed a good bed in which to lie, we shall have a good crop. But the tallest stack of the finest hay would not be as sweet to me as the thought that next year, the gulls might find our furrows worthy of their attention. We need their seal of approval.

I must confess to having committed an act of gross lunacy when sowing my red clover seed. This was the result of casually scanning an aged farming textbook which stated that clover should be sown while the moon is waxing strongly, but does not explain why.

Now, I know there is a branch of agriculture that plans its sowing and reaping so it is in tune with the lunar and planetary movements, but I have not fallen under their cosmic influence. But I remember a chance remark by my old neighbour, Will. He grew onions of shameless proportions and when asked how he did it would only mutter: "There's nowt special abu't onions." But once, in an unguarded moment, he let slip that he would no more dream of sowing onion seed on a waning moon than he would of putting his spade away without polishing it. This gem lay deep in my mind until disturbed by the reference in my textbook, and the now unsettling glare of the half moon through the kitchen window.

With not a day to lose, horses were roused earlier than usual next morning, fed and collared, hooked on to our cultivator and, bleary-eyed, set to work the ploughed soil into a seedbed. Up and down the smaller clods, then breaking them with harrows into bite-sized chunks, until our ribbed roller did the final crumble and left a seedbed as smooth and inviting as silk sheets at Claridge's.

None of this had gone unnoticed — nothing does in the country. Most spectators gawp and coo at the uncommon sight of horses at work, but the really interesting ones are those who watch and say nothing. One old boy in particular caught my eye. He bided his time and then let slip that he worked on this very farm for thirty years. This was what I had been waiting for: someone to unlock the secrets of the old farm.

Kenny remembered having seven Suffolk Punches in the

stable where we now have three; he explained that a hole in the wall was for sweeping old hay out of the loft, and showed me how the yard was sure to flood in a deluge of rain. He remembered the spot where they buried the stock after an outbreak of swine fever. His old farming instincts rapidly returning, he scolded me for leaving bits of bale-string lying about the yard. He talked about the coming of the first tractor, and how the farmer never understood it. On the first day out, he got to the hedge and shouted, "Whooah!" but the old Fordson chugged on, the old farmer shouting "who-oooooah!" ever louder as the ditch grew nearer.

But there was one gap in his knowledge. I wanted to know the names of our fields. Most farmers gave names to their fields, and on larger farms it remains the only reliable way of giving directions. I know fields with such romantic and enigmatic names as Weeping Hills, the Scuts, Glebe Field and Plackett's Walk. But, apart from Stackyard Field, where we were standing, Kenny knew no other names. But he remembered where the hard bits of clay lay, and where the best soil was to be found. Every inch of every field had been ingrained in his mind during those thirty years, but names he could not recall, so we shall have to invent our own.

I am inclined to name one of our fields "P Field", in memory of an accident remembered by another of our aged visitors. As a boy, he was working in this particular field and his truculent horse, complete with laden cart, decided to lie down. No amount of cursing or dragging would get it to its feet. Now, there is a guaranteed way of getting a horse on its feet, and that is to pour water in its ear. But having no ready supply, the lad was quick-witted enough to drop his trousers and perform the trick with that which the Lord had provided. The horse responded. Such ingenuity is worthy of reward, and while future historians will assume we grew peas in our P Field, we shall know different.

Before Kenny went, I invited him to return on a working day and feel the plough in his hands once again. His face lit up. "It's nice to see someone interested in our old ways . . ." then he interrupted himself. "Pit Field . . . that over there was called Pit Field."

I led him over to the clover, hoping for a passing compliment on the texture of the seedbed. I told him I'd managed to sow the red clover seed with several days to go before the moon was full. But he didn't reply — probably thought I was some kind of lunatic.

As if in preparation for some holocaust, the animals have been arriving on the farm, two by two by two. If it turns out to be the prelude to a flood I shall be only too pleased, for the unseasonable baking heat has put me in a biblical frame of mind. In moments of extremely sweaty exhaustion I have imagined myself as a poor figure in a remake of *The Ten Commandments*. Imagine the cruel sun burning down on us, the horses plodding their weary way across the over-cooked soil, hooves raising clouds of dust, and the iron, spiked harrows only bouncing over the clods of soil baked as hard as iron. The normally soggy ditches are arid and awash only with flies and thirsty mice. To water our panting sheep I have carried pails of water scooped from an ever shallower pond and stumbled over craggy clods of earth, spilling as I went.

I felt there were only two solutions to our situation: either Archimedes was going to invent the Screw and save us from drought, or a plague of locusts was going to finish us off. However, the weatherman tells me that he cannot rule out a shower or two in the coming week, so I will revert to my Noah frame of mind, and gather the stock together.

Stock are vital to any self-sustaining system of farming. By grazing the land, not only do they give us meat and fleece or leather, but they bequeath us a hefty lump of natural fertiliser to enrich more soil, to grow lusher feed for more stock to graze. I shall not be happy till we have a rich, steaming manure heap of our own and we are ploughing it back into the land to make it alive and fertile again.

Alice is doing her best to make my dream come true, but one pig alone cannot be expected to take the entire task on to even her ample shoulders. Alice has been with us since December, the result of my wife asking Father Christmas for something "expensive, black and sexy". She got Alice, who is all three: a sow of the Large Black variety who now lives in our orchard, and from a distance could be mistaken for a cannonball on the move. She too has found the heat too much to bear, and has retired for most of the day to her sty. She only comes out for a stroll in the cool moonlight; then she resembles an obese witch's cat on the prowl. However, she can be forgiven her eccentricities, for she has recently lost her maidenhood to a Large Black boar from Bury St Edmunds, and motherhood is clearly on her mind.

The cows, on the other hand, have nothing on their minds whatsoever. They are young heifers, fresh from their mothers and enjoying the closing days of their girlhood: the bull is booked for mid-June. If cows could giggle, ours could match any silly bunch of schoolgirls you could name. I take them an occasional bucket of oats as a gesture of friendship, and for a while they will play it cool and sophisticated, eye me across the meadow, and with a flash of their devastating eyelashes they will eventually saunter across. A couple of yards from the bucket they falter in their step, a giggle breaks out, up go their tails, their eyes take on a mad stare and in a wild frenzy they gallop away. Then they start again. We tried to move them to another field last week, but it was a waste of time.

The sheep, however, have been no trouble at all. But I suspect that this small flock has tasted rather more of the comforts of home life than most sheep do. As we loaded them into the lorry, their sad owner wished each sheep goodbye and, looking one ewe in the eye, said: "Yes, I can see your mother in you, and your grandmother as well." I felt like a kidnapper.

So, for the time being, we have a full house of cows, pigs and sheep ready to take whatever the elements may throw at us. Unlike Noah, we would welcome a deluge: the grass is getting shorter by the day, and until it rains it will grow no more. I was watching the sky the other day above the hillock behind the house, and gleefully thought I saw a dark, heavy thundercloud approaching. But it was only Alice, wandering again.

I am rapidly learning that farming is not unlike the greetings card industry; whatever the calendar might say, you must ignore it and think unseasonably ahead. Therefore on this spring bank holiday I shall be entirely preoccupied with our animals' Christmas dinner. I am determined that our stock will have a good supply of succulent winter feed and so, like a frenzied chamber-maid, I have been getting ready the beds in which the seed will lie. We are going to grow turnips for the sheep, kale for the cows and mangel-wurzels for the lot of them, including horses and pigs. Kale is like a big leafy cabbage, but with no heart. Mangel-wurzels, on the other hand, are all heart: so full of goodness it's a wonder they don't burst. They are red and bulbous, grow sweeter the longer they are stored and have a certain mystery about them. One farmer I know, perplexed by his cows' addiction to them, said: "Mangels, they're ninety per cent water but that must be damned good water."

Preparing seedbeds is pleasant work when you farm with carthorses. Pulling harrows and heavy rollers is good, steady, regular exercise for them; for me it is a satisfying process of reducing the boulders of soil that the plough has left into a fine (but not *so* fine) powdered state, so that the seed will snuggle down into it and be fed and watered as it grows. If the soil is not worked enough, the seed will sit between the lumps and starve to death and Christmas dinner in the farmyard will be a poor affair of hay and miserable manufactured concentrates. I am sure that concentrated feed, like some dreary breakfast cereal, has got all the right proteins, vitamins and minerals in it; but to me it looks too much like astronaut food. As much of the meat produced in this country goes into junk food production, feeding it overprocessed junk in the first place seems to be starting the downward spiral unnecessarily early.

I was beginning to think that I knew every inch of the field in which the kale and turnips are to grow, but every time I take a horse across it there is a surprise. Sometimes it will be an old horseshoe, or a rusty part of an old plough. So far, we have collected an average of five horseshoes to the acre; a farrier tells me most of them are a hundred or more years old.

Harrowing provides good time for thought. It hardly requires any concentration, unlike ploughing, which demands precision. A good pair of horses soon learn the job, and know which way they are to turn when they get to the end of the field. Of course, you must not let your thoughts wander too far, I am sure that my horses sense when I am not concentrating. I have noticed that if I allow my mind to go completely blank, the horses will invariably stop dead and one will turn his head right round to see if I am still behind him.

But nothing is predictable when you farm with horses. The other morning I harnessed Punch, an experienced and versatile horse. I sensed something was wrong. I led him up the field and harnessed him to the ribbed roller. I told him to "G'r up", and he eased one step forward and froze. "G'rrrr *up*," I growled and he tried hard to obey but could not bring himself to do so. Instead, he reared his head and behaved for all the world like a horse that had just had the fright of his life. Rather than risk damage, I led him back to the stable. He breathed heavily and unhappily, I gave him some hay and let him munch. An hour later he was back on the field, plodding merrily along as if nothing had happened.

I get my moods, too. I look at the expanse of the fields and then at the pitiful narrowness of the furrow and I wonder how I will ever get it all done. And then into my mind comes a phrase from an old farmer neighbour: "There's nothing to farming, boy," he told me, "as long as you get on with it."

With that in mind, I am trudging on, making the beds for the kale and the turnips. Our sheep, cows and pigs are behaving terribly well considering they are in the charge of a novice. The least I can do is ensure them a good Christmas dinner.

Late one evening the telephone in our lonely farmhouse rang. The air was still and the peace had tempted the bats out of the barn earlier than usual. High in a tree, our resident owl was warming up for a night's hooting. At the modern warbling of the telephone I dropped the muck-fork and sprinted from the farmyard to the house. "Hello," I gasped. If Hitchcock had been directing the scene that followed, he would have started the murmur of sinister music at this point.

"Mr Heiney? I think I have . . ." the elderly voice quavered, "I think I have found you a horse-drawn binder," he blurted out, like a man who had been trying hard to keep a deadly secret, and had just failed. "I can't tell you where it is, but if you like I will take you there." He paused. "I have to warn you, it is a bit primitive." We agreed to meet at his cottage the following day, and, with a sense of adventure and conspiracy coursing through me, I went to bed.

To explain: it is all very well to try to work a farm using horses, but where do you get the tools? Horses are simple to buy, but horse-drawn ploughs, rakes, mowers and hoes are not so easily come by. Those that have survived are either found rusting, beyond salvation, in the bottom of ditches, or else they have been tarted up and wastefully strewn around carparks. I have sunk to negotiating a bridle off the wall of a tea-shop, but it is hardly a sound basis on which to equip a serious farming venture. So I rely on agents and spies to do the hunting. Every so often one of them will unearth a crock of gold: in this case, a binder.

A binder is a vital piece of equipment. Without it you cannot harvest your corn. It is an intricate and apparently incoherent assembly of wheels, pulleys, cogs and gears and is usually drawn by three horses. It cuts the corn, gathers it together, ties it in bundles and spits them out as sheaves. When the sheaves are stooked, or leaned upright against each other in sixes, you are left with a field out of any heavy Victorian oil painting called *Harvest Scene*.

My informant, a Mr Sly, is a retired dealer in farm machinery and has an encyclopaedic knowledge of the contents of every barn in coastal Suffolk. He promised, however, that this barn would be special. We sped down isolated winding lanes into what he called "bow and arrow" country, although we were never more than a few miles from the A12. We drew up outside a rambling, decayed Victorian farmhouse obscured by dense and undisciplined woodland coming almost up to the bedroom windows. Daylight showed through gaps in the roof, and, in the fields around, the hedges stood as high as trees.

We fought our way to the back door and knocked. It opened an inch or two to reveal a stone sink fed by a lead pipe. Then the door opened fully and there stood Mr Palmer, eighty-seven years old and bright of eye. He was wearing a heavy blue overcoat which, in its long life, had been ripped and carefully sewn together again, but with string.

"Do you want to cut with a binder, do you?" he enquired,

fixing me with a watery eye. "You want to go backwards?" I told him I believed that going backwards was the only way forwards. He said he'd think about it. I sensed that he was wondering if I was going senile. I tried to glean from him some enthusiasm for his farming days, but his mind was set on life's pleasures rather than toils. "Clacton," he said, "Grand place, Clacton. I had some good times in Clacton. "Do you know Clacton?"

We discussed Clacton as quickly as diplomacy allowed, and then I mentioned the binder again. "It's over there," he said. I looked where he was pointing, but saw only crumbling barns and the tumbling bricks of the old granary.

"No, over there!" he insisted, and I followed the line of his stick until I spotted a few roof-tiles above the jungle of vegetation. I looked at my native guide. In his hands had appeared a machete, an oilcan and an adjustable spanner, and in his eyes a gleam of triumph.

A chilly breeze sprang up and I sensed many adventures yet to come before I had sight, let alone possession, of the treasure.

Mr Sly bent down and started hacking at the impenetrable brambles. "Mind where you go," Mr Palmer urged. "I think there used to be a pond under there." More hacking. Mr Sly's machete swung with a determined rhythm. "Is the farm set aside?" I asked, assuming that 100 acres of land lying fallow might as well benefit its owner to the tune of £70 an acre under the government scheme to take land out of production.

"Well," said Mr Sly after a little thought. "It is set aside. But of its own accord, if you see what I mean." We were now in sight of the binder, and Mr Sly put a comforting hand on my shoulder. "Don't be too worried if it looks like a load of old bedsteads," he said. "We'll get it going." He opened his bag, which contained a brown boiler-suit, an oilcan, three spanners and a half-bottle of lemonade.

The warning had been timely. When I finally got full sight of the horse-drawn binder my emotions wavered. I could not decide whether I had unearthed a treasure or stumbled across a scrapheap. It was rusty and filthy, but bone-dry, having been well covered from the rain. The wood was not rotten and, despite the corrosion, when you applied the spanner, bits of it did start to revolve.

I warmed to it. Under Mr Sly's expert control, the oilcan was gushing lubricant into a thousand old bearings. In a burst of excitement he extracted a bit of twisted rod from a heap and declared: "Look, you've even got a spare sheaf-carrier foot-pedal with it." Joy.

If you have never seen a binder, there is nothing I can say to give a true picture of its complexity. But, in principle, it is drawn forward by three horses through a field of standing corn. A knife slices through the crop and a set of revolving wooden sails throws the cut corn on to a moving canvas platform. It is carried along, and then mysteriously upwards until it falls into the jaws of a "trusser" and is bundled into a tight parcel. When the machine senses the bundle is large enough, it throws a string around it, ties a knot, and expels it on to the ground as a sheaf of corn. To be able to translate the steady forward plod of a carthorse into such varied and useful mechanical directions is clearly the work of a genius. Indeed, the man who invented the "knotter" found that his own invention was beyond his comprehension, and killed himself. Watching Mr Sly conducting his symphony of lubrication, I could see how one might easily lose one's grip. After a difficult rebirth, due to the barn roof having dropped a foot since the binder was last used forty years ago, the machine was loaded on to a trailer and brought home.

It is my binder now: last used, according to Mr Palmer, in the year that I was born. I often go and sit on it, and look, and marvel. I don't see the rust; I have in my mind acres of swaying oats and barley, and our Suffolk Punches drawing my binder through the golden crop.

But harvesting with this machine is not simply a fanciful nostalgic exercise. For a start, we shall have long, undamaged straw, which will make comfortable winter beds for stock, or will thatch roofs: the straw that comes out of a combine harvester is smashed and mangled. We shall keep our weeds under control as well. A combine throws out unwanted seeds; putting weed seed back on to the land is the last thing you want if you are not using chemicals.

All these thoughts float pleasingly through my mind as I sit astride the dormant binder. And then a darker one occurs. "I wonder if it works?"

CHAPTER TWO

New Life

Spring 1990

When I woke from a deep sleep with the sound of bells in my ears, I assumed the strain of farmwork was beginning to show. True, we had had a busy couple of weeks, killing weeds amid the turnips and the kale. With chemicals it would have been done in a day, but I prefer a horse-drawn hoe. It is effective and pollutes nothing, being no more than a blade that the horse drags between the rows of plants to chop off the weeds just below the surface.

The snag is that it involves two men (or one man and a deeply reluctant wife) and a lot of walking. If you are the unlucky one who gets to lead the horse, you are effectively standing next to a perspiring ten-kilowatt radiator, you leave the field sodden with your own sweat and a few gallons of the horse's. So I assumed, hearing bells in the night, that the hoeing had drained me.

Then the chime rang out again. It was half past two. Wide awake now, I flew to the bedroom window. In the moonlight I could just make out the pregnant shape of Alice, our Large Black pig, making frenzied music with her feeding bowls.

Pig troughs are no lightweight affairs: they are cast-iron

rings which it takes two men to lift. But Alice has been blessed with a power-packed snout, and it is nothing for her to slide her muzzle under one of these hefty troughs and, with a flick of her head, heave it in the air. When it comes down to earth, spinning, it sounds like the very bells of hell. From the bedroom window I loudly advised Alice to cut out the Quasimodo impersonation, and went back to sleep.

Of course, pregnancy does funny things to women, and pigs. Next morning I found that as well as revising her dining arrangements, she had also done a thorough spring-clean of the sty, moving the clean straw out into the sun and leaving the grubby stuff in a heap near the spot where she dungs. "Daft old pig," I muttered into her floppy black ear, pouring her breakfast into the relocated trough.

A few hours later, we had eleven piglets. First there was nothing, and then in no time at all there were eleven shiny black squealing creatures that slid from their mother with the greatest of ease, shrugged off their cling-film and staggered in the direction of a nipple with a determination that brought a tear to my eye. It all took place nonchalantly, out in the sunshine on the clean straw. There was no fuss, except what I made myself as I ran to tell the children. "There's two!" I cried. Then ran back to the sty. Then back to the house. "Three!" I sprinted from farmyard to house bringing news of the births. By the end I was bursting with pride and panting more than Alice.

I rang the owner of the boar to tell him the good news, and he was delighted. I remembered picking her up after she had been six weeks on his farm, and not knowing quite how to phrase the question which would elicit from him whether or not a mating had taken place. “How have things, er, been?” I enquired. He considered. “I’d say he’d stocked her well, my old boy. Yep. Stocked her well, he has.”

As soon as she was home I marked the calendar. Pigs have a convenient gestation period of three months, three weeks and three days. We now know that the happy union took place on her second night. It’s lucky that black pigs can’t blush.

Some might consider it bad pig management to have been taken so much by surprise by the birth, but I had been relying on the advice of an elderly neighbour. He had been positive: “She won’t be havin’ them little ’uns yet. Look. She ain’t appled-up.” He pointed to her udder. “Yer know what I mean?” he asked, and cupped his hands. “Appled-up. She ain’t appled-up yet.” She never was. Hence my failure to interpret her musical, midnight nest-building session.

She did not need me, anyway, that day. There was a brief crisis when one piglet got caught beneath its mother’s bulk as she turned. I was tempted to dive in and help, but as soon as the little one shrieked, Alice rolled the other way. It was the only moment when I thought I might have to play midwife, which was just as well as I had been rather dreading it, ever since I’d read a 1920s book which said: “There are few problems in farrowing that cannot be solved by good humour and a plentiful supply of lard.”

But we needed neither. Alice did it her way, unaided and with great dignity. She has done us proud. Let the bells ring out.

I shall throttle the next person who says, “Remember, make hay while the sun shines.” I am only too well aware that we had the hottest April and May since the earth was a molten inferno, and I admit that I recklessly wasted this blistering interlude in parading around county shows, fancying myself outside the Country Landowners' marquee and strutting up and down the lines of animals pretending to know my dairy shorthorns from my Aberdeen Angus. That was when we should have made the hay.

Now that it is raining I am a picture of sorrow and contrition. But not as sorry a sight as my newly mown grass, which under the influence of the departed sun and wind would by now have been transformed into succulent hay. Drenched, it is turning into black, slimy goo. I have learnt my solemn lesson: what is missed can never be retrieved.

Farming is the fullest of occupations. If not keeping your hands busy it keeps your mind spinning. I now find my nights are broken at the sound of the slightest drop of rain, or a cough from an animal in a far field. For months I have not been able to sit down to a meal without an eye wandering towards the expectant sow in her sty. But nothing so far has been as agonising as this apparently simple business of cutting grass and making hay. I have been offered ten acres of grass some miles away, and since it is too far to walk horses and machinery safely, the hay is being made by my good neighbour Mr White, who fortunately has the patience of a midwife dealing with a father at the birth of his first. We speak on the telephone every morning and brood over whether the hay might be ready to turn, or leave another day? Or, try it later this afternoon? Or not at all? My wife, overhearing these protracted conversations, says that multinational empires have been traded with less discussion.

When not on the phone to Mr White, I'm glued to the weather forecast, or the weathervane on the stable roof. "It won't rain as long as the wind is south-east," I decide. Then I ring Mr White with my thoughts. Poor man.

At home, our modest haymaking is part of a waste-saving experiment. Our next door neighbour, though by our standards an intensive farmer, has a deep streak of conservation running through him. Round the edges of his fields he has sown grass roadways several yards wide from which the countryside gets huge benefits at little cost. For a start, his weedkilling sprays do not reach the verges and destroy the wild flowers, the grass stops the hedgerow weeds creeping into the crops, and apparently birds and animals enjoy a saunter down the grassy promenade to dry off after heavy rain. Rather than mow the grass and let it go to waste, he invited me to make it into hay. I'm happy, he's happy and the wildlife is throwing a party.

In the non-intensive way we intend to farm, the "waste not, want not" philosophy runs deep, not only in the obvious recycling of animal fodder into dung into compost and back onto the land, but in almost every aspect of our farming it turns out that what intensive agriculture considers waste, we have a use for. Chaff, for example. Chaff is the husk that envelops the grain of corn. As part of the modern harvesting process it is removed and discarded. Our antique harvesting methods, involving a binder and a threshing machine, preserve the valuable chaff and we shall have huge, comforting bags of it to see the animals through next winter. The small building next to the stable was a "chaff house", and it will be again.

We try to observe wartime disciplines. No crust of bread

is thrown away, but cherished as if rationed and put in a bucket at the back door to be boiled up for the pig. Even when we rinse the milk bottles I try to remember to add the washings to give a bit of variety, and butterfat, to the pudding. But I have to be careful to avoid any illegal adding of meat products and now find I am reading the E numbers on processed foods in the pig's interests, rather than the children's.

Many tips I have learnt from musty farming books written during the Second World War to encourage farmers to produce more from the land. One suggestion which particularly stuck in my mind: "A farmer visiting a town with his truck should never return empty but rather bring with him a healthy load of sewage sludge from the town's works." Not perhaps, when the sun shines.

Wild geese are not the only creatures which can lead a man on a hopeless, heartbreaking chase. Since I took on this farm almost every animal has shown equal talent. Animals are not a problem when they are contentedly munching their way across the landscape; the trouble arises when they have to be moved to another part of it and do not want to go.

There is often no choice. In our case, the heifers had to be put on to fresh grass or they would starve; the sheep had to be robbed of their sweltering fleece or they would melt. Then there were the chickens.

We were given a bantam hen and chicks, and sternly warned to cull the cockerels as soon as puberty struck: three randy young bantam cocks pursuing a couple of maiden hens are not conducive to a peaceful farmyard. Not having the skill or the inclination to wring a chicken's neck, I built a wire-netting run and put them in it until an executioner could be found. Within minutes an escape plan had been hatched: the boys were under the wire, free and crowing in defiance. I have now decided on a new approach to poultry keeping. I shall fence in the vegetables instead, since they are slower on their feet, and let the chickens have their freedom until fat enough for the pot. Just how I shall strike I do not know, but my new farming motto is: never admit failure, call it a change of policy.

The heifers are a more serious business. They are three young Red Poll cows with a prize-winning pedigree worthy of *Debrett*, and for financial as well as protocol reasons they deserve royal treatment. Being of an old-fashioned breed, they will make do on meagre rations. However, I decided that simply making do on arid grass was not good enough for them, and that they must go to pastures new.

When they first arrived the cows were wild enough for a western rodeo. An outstretched hand had them galloping away in fright, a muttered word in their silky, red ears made their eyes roll in terror. But not any longer. The girls have succumbed to my charms. I have learnt two things about cows: that they are curious, and that they are anybody's for a bucketful of oats.

Every morning for a week I rattled the bucket, let them get the scent of oats, and stood still. Day one got no response. By day three they were within an inch, by day seven we had made friends. After that I built a pen out of rusty old gates in the corner of the field, backed the lorry in and the girls ambled up the ramp, as happy to be on the road as a load of children on a school outing. No change of policy needed there.

The sheep, however, are a different matter. Our small flock lives on a grazing marsh which is known for its wildlife. I would care to bet, however, that nothing on this isolated wetland is as wild as our flock of young sheep. Despite the conquest of the cows, I am beginning to think that having so much youthful stock is one of the main problems of starting a farm. Every animal is going through its teenage delinquency at a time when the poor fledgling farmer really needs mature, stable, motherly beasts around him.

Anticipating the problem of catching sheep without a dog, last Christmas we bought an orphan lamb. The idea was to raise it on the bottle, make a pet of it until it believed it was human and would come when called. Once the lamb had been returned to a flock, we would only have to go down to the marsh, call its name, and it would come to us with the rest of the flock following in line as sheep do. We called our ewe lamb Shambles. This was prophetic.

Six weeks after she had been liberated, we went down to the marsh and called "Shambles!" Disturbed birds took flight, but not one sheep's head raised itself from the grazing position. "Shambles!" I shouted again, loud enough to stir the rabbits this time. Not a flicker. Then we made a fatal mistake: we decided to try to round up the flock ourselves.

I had with me a broad-chested chap who has Olympic aspirations and could be said to be "in training", and an elderly marshman, well past retirement. I offered to get behind the flock and edge them forward while the other two steered them in the direction of the gate. When I banged my stick lightly on the ground, the flock fled as if I had fired a starting gun. The athlete advanced with arms and stick outstretched to head them off, a human barrier. The bleating horde jumped, one by one, over his arm. He swore. They were heading for the marshman now. "I was in the war," he shouted, readying himself for the battle. "Gallipoli, I was at." The enemy charged, jinked around him, and advanced victorious towards the horizon, the traitorous Shambles leading the column. "I'll head 'em off," the old boy shouted and using his detailed knowledge of the marsh, shot into the bracken like a stormtrooper.

No sooner was he into the undergrowth than the sheep were out the other side, hell bent on inflicting further humiliation on the athlete. They were panting by now, but not half as much as we were. We gave up. In a mere thirty minutes, a small flock of sheep had got the upper hand of their alleged master, a Desert Rat and an Olympic hopeful. Remembering that all problems can be solved by the adoption of a new policy, I have reached a decision: this farm is going to have a sheepdog. I have reached another decision. It will not be a young one.

My farming week started far from home, in the bleak Norfolk Fens: a low-lying, fertile tract of land where the wind always blows chill. It must be a lonely life for a Fenland farmer: impassable drainage ditches make every man an island, and neighbouring farms can be miles apart via the nearest bridge. I have always imagined Fenlanders as being dour and remote.

My host was an amiable sort, however, not at all deranged by his isolated existence in a bungalow built into the breach of an old sea-wall. A few miles to the north were the swirling waters of the Wash, and the constant hum of pumps suggested that we were some feet below sea-level. I avoided mentioning the greenhouse effect.

We had a cup of tea and, just as I was reproaching myself for thinking Fenlanders any different from the rest of us, he picked up an aged carving knife, held it a couple of inches from my nose until the steel was too close to focus on and, with his eyes wide, said: "This was my father's. He used this to kill pigs. Kill pigs!" Then he pretended to slash his throat, impersonating the bark of a dying swine as the steel flashed across his neck. I finished my tea, hurriedly.

I had done my deal and secured my treasure: a horse-drawn swath turner, which I towed home to Suffolk rejoicing. This machine flips heaps of newly mown grass so that it can dry in the sun and become hay. I have long coveted a machine I once saw which did this wonderfully: it had a series of mechanical flippers which kicked the hay high in the sky as the horse walked along, and resembled a robot attempting the Charleston. The machine I had just bought was more dour and Fen-like in its action, but just as effective.

I arrived home to be told by my wife that she had managed to spend twenty minutes lost in the middle of a three-acre field. This is quite an achievement and so, with some curiosity, I followed her insistent directions. They led to a patch of land which has had a question mark over it: some of our land is still carrying the crops of the previous owner and I thought he was tending this field, while he believed that I was. The result is that nature seized her opportunity: the weeds have thrived on last year's nitrogen residues, and we have stalks of mayweed and thistles that reach higher than a kilted Scotsman would find comfortable.

I stood wonderingly in this lost world, then came a puff

of wind through the jungle. From each of the thousands of thistles drifted a handful of fluffy, innocent-looking thistledown. I froze. A million seeds had just taken to the air, and away with them had drifted our hopes of a weed-free farm next year. Without chemicals, organic farmers have to give each weed personal attention.

Watching the thistledown in flight, it was as if I had just seen the entire crowd at Wembley Stadium rise, and knew that next spring I would have to shake hands with each of them.

One skirmish with weeds has already been lost. You may remember that the first crop I sowed was clover, which I did while the moon was waxing, on the ancient theory that the growth of the moon encourages the budding of the crop. So it did: a good crop of clover appeared, but so did numberless weeds. We had docks, fat-hen, mayweed, poppies and yet more thistles. Despair. I announced at breakfast that I was off to kill the fat-hen and could not understand why the children's eyes filled with tears. When I explained that the victim was not our clucking, speckled friend, good humour was restored.

I need not have fretted. When I turned to yet another of my musty textbooks I found that the invasive fat-hen weed gets rave reviews as "an indicator of high soil fertility"; that mayweed's "profusion of leaves makes a valuable contribution to soil fertility if mown", and that when the author got a similarly disastrous clover crop he simply mowed it and let the mowings lie until they rotted and fertilised the soil. Twelve months later, after a little carefully controlled grazing by cattle, he had "a field that was without superior in Britain". It seemed worth a try.

If we can have no clover hay this year, we can at least have hope.

Out came the horse-drawn mower, and one man and his horses went to mow a meadow. The blade clattered through the growth, reminding me uncomfortably of the Fenlander's pig-sticking knife. I began to view the field as an exciting experiment in natural fertility, rather than a beginner's failure.

I now gaze upon my brownish field of dry stalks in certain expectation of the finest sward in all England; even the fine haze of thistledown wafting down the hill cannot depress me. Much.

CHAPTER THREE

Heatwave

Summer 1990

The heat is on, and it is not entirely due to the weather. Temperatures have been soaring not only on the parched land, but in the deepest recesses of all our souls: the animals included.

Carthorses soon get steamed up in sultry air. Flies don't give them a minute's peace, and they are forever violently swishing their tails, stamping their feet and twitching. The result is that the poor horse finds himself under attack on two flanks: from biting flies or from his master for not standing still. The ability of a carthorse to stand like a statue when ordered is second only in importance to his talent for walking, unguided, in a straight line. When the heat starts to rise, it all goes to pot and work becomes nearly impossible.

I become a handful too. I cannot bear the sensation of rivers of sweat in which the midges paddle. I have been out with my hoe in the mangel-wurzels, praying for the chill winds to blow so that we can all get back into overcoats. When it all becomes too much, the horses and I retire to the stable where the air is always cool and damp. I am often asked why we keep the horses inside during the day and only let them out to graze at night. The answer is simple: the horses like it that way. If I were to put them in the meadow by day, they would only stand by the gate pleading to come back inside. If impatience got the better of them they might lean against it, and be through in an instant. Carthorses have a habit of voting with their feet.

Alice, the Large Black sow, has been voting with her snout. You will remember that a few weeks ago she was delivered of eleven lively piglets. Well, all eleven are thriving, squealing, biting each other's ears, sleeping a lot and spending long, blissful feeds glued to their mother's nipples. But for Alice, the novelty of motherhood is beginning to wear off. When she is tired of their attentions she flops on her belly so that her ample stomach shrouds her udder. It's her way of saying, "Go off and play, dears." After a spell in the hot concrete sty, it didn't need an agony aunt to advise a change of scene for the young mother. I decided the sow and family should go to the orchard.

Pig-moving is a game of diplomacy. You suggest a direction in which she might like to go, and hope she takes the hint. There is no point in prodding with a stick, for she will freeze. The game needs as many people as you can muster, each of whom carries a board: if a pig cannot see a way ahead, it will not go. You use the boards to deflect her progress: if she heads the wrong way, stop her with a board and let her see only in the direction in which you would like her to go. She retains, of course, the option of standing stock still whatever you do with the boards, but let us draw a veil over that.

At pig-moving time, any visitor is in danger of being

pressed into service. It was unfortunate for our friend, the art dealer, that he happened to call that afternoon. Italian leather shoes that had only known the gentle caress of a Bond Street pavement now found themselves up to the buckles in sodden pig litter. But pig-shifting brings out the best in people: rather to our surprise, he entered into the spirit of the thing and when the moment came to round up the piglets he slithered and pounced like a professional swineherd. Fingers that only hours before had been stroking gilded frames, grabbed the hind legs of the protesting, wriggling creatures. When he next raises a finger to bid at a Sotheby's auction, few will suspect where it has been.

Alice and family love the orchard. She places her ample rear against the shakiest of the old apple trees and wriggles her behind till the young apples cascade onto her waiting piglets. They have even made themselves a mud-wallow and are as happy as a family on Blackpool beach.

So why don't our heifers go and roll in some? They have other things on their immature, feminine minds — like the boys next door. In the field next to where they have grazed undisturbed for some weeks, a herd of young stock appeared. I knew nothing about it till the phone rang just before seven on Sunday morning — "Mr Heiney, there's a problem with your cows!" I felt like Mr Barrett of Wimpole Street, discovering that my girls had been out on the razzle. The heifers were, as we delicately call it round here, "in stock". On heat, in the heat. Overexcited, to a degree. We herded them back into the field where they should be, reconnected the electric fence, and turned our backs for a moment. This was long enough for them to toss aside the wire, which was pulsating with 5,000 volts, barge through a spiky blackthorn hedge and dive through three strands of barbed wire. Very perplexing. Especially as my keen farmer's eye had by now detected what they had not, that all the animals in the field next door are girls, too. I blame the heat. It is unsettling us all.

How did your holiday packing go? Mine was frenzied. I realise that with harvest coming slap in the middle of the holiday season, farming and summer breaks are going to be unhappy partners from now on. But as we have no corn crop this year, we thought the chance of a final summer fling was worth taking. Next year it will be binders, sheaves, pitchforks and sweat. This year it was to be Ireland, open seas and green hills.

Before leaving, I dashed down to the marsh and judged that the sheep had enough nibbling for the short time we would be away. Not so the cows: in a pleading phone call I begged a meadow off a neighbour and moved the three heifers on to that. The pig family was given sanctuary in the orchard — Alice's endless supply of Golden Delicious manna from heaven. Having created a picture of rural contentment and organised a feeding rota, we closed the gate behind us and headed west.

I envy the farmers of western Ireland two things. Firstly, they have grass of a succulent greenness that I have not seen at home for many months. This peaty abundance explains why the Irish produce so much milk with so little effort. A wild rumour spread through the village in which we were staying that a hero from the local creamery had discovered a way of turning milk into alcohol. If it had been true, St Patrick's patronage of the country might well have been under threat.

But what I most envy the Cork and Kerry farmers is their attitude. It is not that they don't care, it is simply that they only expend worry on things which are really worth worrying about. And wandering animals do not come into that category.

Take the herd of cows that strolled past us as we sat on a stone wall in the middle of nowhere. At home, cows on the road would have an escort of men in pick-up trucks and a tractor with lights blazing and horn sounding. Out in the rocky wilderness no man was to be seen; just cows, shuffling and sniffing the hedgerows and eyeing us with suspicion.

Ten minutes after they had passed, an ambling figure appeared, shabbily dressed and with a wind-burned face that had spent many happy years gazing into glasses of stout. "Have you sheen any cowsh?" he asked in a half-hearted way.

"Yes," we said, "heading that way," and we pointed up the hill.

"Oh, dear, oh dear, oh dear," grumbled the farmer. "They know they're not shupposed to go that way. Oh dear, oh dear, oh dear." Then, to our surprise, instead of heading after them, he turned and went the other way saying, "If yer shee 'em comin' back, jusht keep 'em pointin' wesht." And he was gone.

Ten minutes passed and his confidence was repaid. Back came the cows wearing a guilty, out-of-bounds look on their faces. They were pointing west, and so we let them by unchecked.

This left a deep impression upon me. When we returned home I vowed that things around here were going to slow to the same enviable pace. No more running after stock, no more palpitations at the sight of a sheep on the wrong side of its fence: less time on my feet and more time on the bar stool.

This calm transcendental state lasted all of half an hour. I went down to the marsh to count the sheep, and found one missing. The chap who had been keeping an eye on them wasn't sure where it was. Shaken, but not stirred, I simply said, "Ah well, she'll be back, she'll be back." In the baking heat of the farmyard, the intoxicating lazy scent of clover hay elevated me on to an even higher plane.

"It looks as though the piglets have been escaping," my wife said.

"Is that so?" I replied. "Well, I hope they had a nice stroll. I'll mend the fence, sometime." The carthorses looked well. I must get you boys back to work, I thought to myself, . . . one of these days. Then the phone rang.

I went through the usual pleasantries: yes, we'd had a lovely time, the weather was terrific; yes, wasn't it hot while we were away? Then the bombshell dropped. "Did you know," said the voice, "that while you were away, your cows got into your neighbour's swimming pool?" The holiday mood vanished. My mind raced with visions of Red Poll heifers doing the butterfly, of hooves jammed in filters and cow-pats on the cow-patio. I realised that the crowded, suburbanised south-east of England was, after all, a far cry from the wild shores of Dunmanus. I worded the insurance claim in my head and wondered if anyone would believe it.

Upon discovering the truth of the matter, which was that the cows had merely got their heads over the fence to nibble the grass round the pool, I felt easier. Still, had things been as bad as I'd thought, I doubt I would have had the courage to look my neighbour in the eye and ask him to "Keep 'em paddling wesht!"

The wildlife of Suffolk is crying out in unison, "Juliet, Juliet, wherefore art thou Juliet?" In this intensive farming area, the wildlife is firmly of the belief that she is one of their few friends. And Juliet Hawkins, young and lovely, returns their devotion. She *is* our Farming and Wildlife Advisory Group, known less romantically as FWAG. Each county has one, paid for with a species-rich mixture of cash: conscience money from the agro-chemical industry, a grant from the Countryside Commission and others, and old-fashioned fund raising. Last weekend, the Suffolk branch held a Hog Roast, the pig presumably happy to make the ultimate sacrifice to help his compatriots down in the green verges.

Juliet Hawkins's job is to move the birds and bees slightly higher up the farmers' list of priorities. Since this is a county where one smallholder was recently asked to restrict the movements of his pet duck as it was threatening the neighbour's corn crop — which totalled no less than 600 acres — I would imagine that being a wildlife adviser here is like cheering for Everton in the middle of the Liverpool crowd.

So I asked her round. Anyway, I thought that having spent every waking minute of the last six months worrying about the soil, what goes into it and what might eventually come out of it, the time was ripe to raise my eyes to a wider world of nature.

Like a green tornado, Juliet Hawkins swept around the farm, her eyes scanning ditches, hedges and verges with the enthusiasm of Patrick Moore discovering a Black Hole and the reverence of Arthur Negus with something chipped from Staffordshire. She dismissed our hedge as being "quite recent" (only a couple of centuries) but was thrilled by our pollarded elms which, she said, were sure to denote ancient boundaries. We had a moment's silence out of respect for our pollarded hornbeam: it is apparently "quite, quite rare".

We walked up the farm to the old meadow, and the thrills came fast and furious as each tuft of rough grass was declared to be home to the most special of butterflies. Pity. I've been promising myself for weeks to tidy that mess.

By the time we had done a complete circuit of the farm, I fell, dizzy and panting, to the ground, exhausted by the sheer intensity of the wildlife. I had just enough breath left in me to start steering the conversation round to what I hoped was going to be a lucrative discussion about how a few pots of gold might drift our way to replace our ripped-out hedges. But there was no peace. Ms Hawkins had caught sight of a huge bird. She declared it to be a marsh harrier; I had thought it was a seagull. I suspect it came to serenade her: "Oh, she doth teach the torches to burn bright!"

The Farming and Wildlife Advisory Group have done great work in this county in persuading farmers that even if you factory-farm, you can always find room for the wild side of life. So persuasive have they been that I am told of one farmer who, remorseful at the destruction of his hedges and woods twenty years ago, is now replanting them all by hand as a penance.

But the public have as much to learn as the farmers. Take my 200-year-old hedge, of which I am rather fond. It is largely spiky blackthorn to dissuade stock from barging through it, and over the years a wealth of wild roses has twined into it. There are brambles too, which mean blackberries. When I asked how best to care for it, I was advised to "cut it down to within four inches of the ground". A conservationist calls it coppicing and can get away with such behaviour; a casual observer might call it vandalism. In fact, Ms Hawkins told me of an old man who, complete with hedger's traditional tools, was doing a splendid job of coppicing a farm hedge. He gave up when too many tourists accused him of blighting the countryside. But he was doing exactly the right thing: the hedge would regrow very quickly to be twice as thick and just as lush in a very few years. If I take the chainsaw to our hedge, it will be under cover of darkness.

I certainly foresee problems with our pond. It nestles in a quiet corner of an old meadow and, though now overgrown, with a little loving care and the help of a great big digger it could become our premier wildlife haven — providing I keep the ducks off it. Ducks, it seems, kill the insects, frogs and toads and erode the banks with their coarse, unselective webbed feet. "Encourage the moorhens, but not the ducks," she warned me. This is all very well, but how do I explain to the uninformed and the children that I'm shooing the pretty little ducks away in the name of nature? Poor ducks. And poor me, for doesn't that put me in the same miserable class as the barley-baron who ordered the lone duck off his land? Problems, problems. Nobody could be keener than me to fill up every inch of his hedges with wildlife, but the course of true love never did run smooth.

I have reduced the unemployment figures by one, and he's

not pleased. At least, that's the impression he gives. Despite enforced idleness staring him in the face, he doesn't appear to be the slightest bit grateful.

He's called Flash; a Border collie with a determined and experienced approach to the sweaty and frustrating business of moving sheep. Neither I nor any other man can hope to match a sheep in speed or cunning. Only a sheepdog can, and that is why Flash is now part of the growing menagerie on our farm.

Word soon got round that I was looking for a dog; and a phone call from a Norfolk shepherd told the tale of a thousand-acre farm that was to be converted into a golf-course, with its flock of 2,000 ewes being sold. They call it diversification: I call it vandalism. Those of us with precious few acres find it galling that others can fling aside huge tracts of land. As I arrived at the shepherd's cottage, I could see from the glum expressions on both their faces that Flash and his master shared my view. Both had been given the boot.

At his master's command, Flash came storming out of his kennel and immediately took a neat fold of flesh from behind my knee and pressed it between his teeth, hard. It was not meant to be affectionate. If this was going to be an enduring relationship between one man and his dog, I felt it could have got off to a more promising start.

The three of us drove in an old Land-rover down winding tracks and through aged woodland (due to be flattened to make a "green") till we came to water meadow (due to be drained to make another "green"). A thousand sheep were grazing idly, until Flash leapt out of the car and crouched in that tensed, concentrated pose that is the hallmark of the trained sheepdog. He was bursting with desire to bring each and every sheep to his master's feet; but he had been given no word of command and would not move an inch until it had been given.

"Come by!" He stormed to the left, keeping far enough from the sheep not to cause any panic. Slowly they edged together as he moved up on them. Another command had him falling to the ground, frozen until yet another brought him to his feet. He was a cracking dog, as Phil Drabble might say. Money changed hands (quite a lot of it, as a trained dog is a valuable animal) and with a hint of reluctance, Flash slid into my car and we made our way home.

That was last Saturday. It was Wednesday before I got anywhere near him again. I tried the soft approach, with lots of cooing and "good boy, good boy" but the bewildered Flash just slunk into the corner of his kennel, bared his teeth, snarled and refused to budge. I can't blame him. For all he knew, I might have been a golf-course developer.

Then a neighbour who is a shepherd had a bright idea. He brought his collie Tess, a bitch who is anybody's for a cuddle and a Bonio biscuit, and put her in with Flash. The change was dramatic. It was as if an imprisoned man found his gaol had been turned into a harem. Now I could boldly venture into the kennel for a pat and cuddle of Tess — and within a day Flash was wanting attention too. By Friday he was licking my hand. I call that progress.

Now, I have to confess that I have been down this path before. Six years ago, for the television series *In at the Deep End*, I was given a dog and instructed in the art of sheepdog trialing. The dog was called Tim and led a frustrated few months, for in those days I had no sheep. The poor animal had to be content with rounding up the only things that happened to be moving in our garden, which were black rubbish sacks drifting in the wind.

By talking to shepherds, I learnt that sheepdogs live hard lives, by pet dog standards. They rarely live in houses. (It was the proud boast of Flash's owner that he had "never been in a house, never!" This was a selling point.) Nor do they seem to expect much in return for a hard day's work: a pat and an affectionate "good dog" is worth more to them than a gold medal at Crufts. But if you have seen, as I have, the dedication and the instinctive skill of dogs working sheep on the wild mountains of Scotland and the rugged hill farms of Wales, you cannot fail to be of the opinion that one good working dog is worth all the pampered poodles in the world.

As I looked at Flash, hoping that he and I would forge that unique bond that exists between shepherd and sheepdog, my mind recalled the lessons of years ago. "Come by" to send him to the left; "Awaaa-y" to move him to the right; "Look back!" if he'd left any sheep behind. His ears pricked up, his head leaned to one side and I saw in his eyes a look of willingness, and, I fondly thought, a hint of devotion. Then, for a joke, I said to Flash, "Golf-course!" He snarled. I think we are going to get on just fine.

Farming is not a single occupation. Looking back over the past six months, I would guess that far less of my time has been engaged in tending plants, cultivating soil and caring for stock than has been spent in activities involving spanners, sledgehammers and shovels. When the water trough needed a new ballcock, I had to be plumber; when we put up a shed for the cattle I masqueraded as architect. I have turned blacksmith when vital bolts have seized, and civil engineer when a short roadway had to be built at the muddy entrance to a field. You don't necessarily have to be competent at any of these professions to be a farmer; you only have to be adequate.

But there is one area where a farmer must never show any inadequacy, and that is in his dealings with other farmers. Not that I have been taken advantage of, as far as I know, but farmers have a natural inclination to seek a quick harvest and if they can reap a few shillings from an uncertain novice some will do so.

To give you an example, I had a phone call the other day from a man who wanted to sell me manure. The first thing I asked was the price. Satisfied with the cost, I went to see the muck. "There's fifty ton there, guvn'r," said the vendor. "Will you take it all?" Now, anyone who had been farming for some time would have known how much muck he was looking at, but for all I knew it might have been five tons or five hundred.

There have been many uncertain moments like that and when such a crisis of knowledge strikes I always turn to a volume roughly the size of a pocket Bible. It is the *Notebook of Agricultural Facts and Figures for Farmers and Farm Students*, 1924, and the compiler has the unlikely name of Primrose McConnell. Until I read the preface, I had assumed this authoritative person to be some fierce lady of the land who could plough like any man and spit nails into horseshoes; but I find that Primrose is, in fact, a gentleman, describing himself as "Yeoman Farmer of Southminster, Essex". In a sad introduction he dedicates the volume to his son, Cpt. McConnell MC, who was killed in action on the Salonica Front, September 1918.

This precious volume holds all secrets. Where else would you discover that sheep dung requires four months to ferment, rising from 141° Fahrenheit to 158° Fahrenheit? Cattle dung needs a lengthy eight months and rises only from a cool 95° to 113°. In my search for a figure that would give me a clue as to the amount of muck in the heap that was on offer, I noticed in passing that "The horse produces 12 tons of manure a year, a cow voids 57 pounds of solids daily." Primrose gives us the chemical compositions of cow's urine, both stale and fresh; this, presumably, is to aid the keen farmer who sits with a collecting bowl. By the way, those in the wealthier shires who are unlucky enough to have an accidental seepage from stable to swimming pool may like to know that horse urine has a specific gravity of 1.06. You'll float in it all right, I think.

The breadth of McConnell's knowledge is astounding. He tells me that in windmills, the millstone revolves five times to every revolution of the sails; at a wind speed of 20 feet per second an average windmill will grind 5 bushels of corn every hour. More importantly, "1½ inches of ice will support a man; 4 inches will carry cavalry and light guns, 5 inches will bear an 84-pound cannon and 18 inches will support a railway train."

It may seem to be a compendium of use only to desperate compilers of quiz questions, but to those of us who farm in a style of which we hope Primrose McConnell would have approved, it is a constant source of sound reference.

How else would I ascertain that "a labourer [usually me] can fill 18 loads of dung into a cart in 8 hours" or that "a man can pitch 4,000 to 5,000 sheaves of corn in a day"? My problem about the stack of manure is solved on page 15 where I am told: "To the area at the bottom add area at eaves; to this add the product obtained by multiplying the sum of the lengths by the sum of the breadths; multiply this by one-sixth of the perpendicular height of the eaves — gives contents of the body." In the end I decided to give the man the money.

However, if you and I should ever be doing business in the future and at some crucial stage in the negotiation I ask to be excused, it will be for no other reason than that I must consult Primrose McConnell. Farmers, of all men, must know exactly how many beans make five.

Slowly, but surely, my dreams of being a small farmer are coming true. I know it is silly, but I always felt that mine would not be a "real" farm until, in my very own stable, a certain scene was re-enacted in the style of *All Creatures Great and Small*. It is the one where the harassed vet calls upon the grumbling, near-bankrupt small farmer. It is usually snowing and the chilly northern welcome goes: "Eh, bah gum, vet-er-in-ary. Tha's too late to save our Daisy. Where's tha bin?"

This is the cue for the vet to sink the entire length of his arm into the rear end of the unfortunate cow. It is a simple and moving scene, which is presumably why they show it nearly every week.

In our case it was a sick carthorse that needed attention. I played the part of the farmer by greeting him with my running grumble, the one about the continuing drought (I threw a radio across the barn last week when they played "Raindrops

Keep Falling on my Head"). The vet, a Welshman, was cast in the heroic Herriot mould. No sooner had he looked at the horse, than he demanded a bucket of hot water.

Buckets of hot water have always been the cue for drama. By the time I had filled the bucket, he had removed his jacket and rolled up a shirtsleeve. Without interrupting the flow of his conversation about the Gulf and the prizes for the breed society's autumn raffle, he delved into the innermost working of my Suffolk Punch with the delight of a child attacking a bran tub. He raked around a little and confirmed a case of worms, and not one stain on his starched shirt.

Two days later worms raised their heads again when my organic farming adviser came to call. With the gravitas of the vet asking for hot water, he requested a spade. We marched out to the far field and dug. You do not find out much about soil by just looking at it. You have to get beneath it, to see the way it holds itself together and how it changes the deeper it goes. You have to develop the critical sensibilities of a common earthworm: worms know worthwhile soil when they feel it. They improve the soil as they burrow: they aerate it and break it up allowing surplus water to drain more easily. To have a field full of active worms is to employ a non-stop workforce in subterranean soil improvement. And they're free. But worms, alas, are killed by some agro-chemicals. So it was more in hope than in expectation that we examined each spadeful like men panning for gold. Not many worms were found. My adviser prescribed a good dose of compost, and rolled his sleeves down again.

The next day it was me who squirmed. I hate, above all, the sight of tractors marching across our land. I associate the rumble of the diesel engine with personal failure and submission to the system. And there they were. I was uneasy all day, even though it was not disillusionment with horses that made me invite them, but rather the reverse. Land that has been worked with heavy machinery for many years needs a period of readjustment before gentler horse-drawn equipment can tackle it. The arrival of my neighbour, the ever-patient Mr White, on his mighty red tractor marked the first stage in its reclamation: the powerful machinery will break the cloddy soil into crumbs, into which the horses and I will plant and roll next year's crop. Gradually, with the return of the worms, the soil will become light enough for horses to work again.

I consulted the *Radio Times* before going out to confer with Mr White: he plans his operations around the radio in the cab of his tractor, and knows exactly how many acres he can plough during *Afternoon Theatre* without missing the end. I chose my moment carefully and took him a cup of tea during the shipping forecast.

As we sipped in silence, neither of us dared to mention the thought that was going through both our minds. Even his heavy machinery had failed to break our drought-stricken land into soil fit for sowing grass seed. The lumps are simply too big. Earth has to wrap itself around seed so it can feed and be watered; without rain from above or worms from below, ours is as inviting as a rock face. He hopped back into the cab for *Woman's Hour*, and I walked gloomily back to the farmyard. Clearly, neither I nor the horse will be entirely happy until the worms have turned.

Soon I shall harness the horses, drag the plough from the

barn and start the long slog of renewing our soil, furrow by furrow. I shall not expect quick results: a fit man, with equally fit horses, should be able to plough one acre a day. He walks eleven miles doing so. It is going to be a long march. But that is next week.

For the moment I am content to lounge in the hay in a newly acquired state of trance, smell the dampness in the wind and reflect on my first six months as a farmer.

It has been a sweat. The learning curve has been precipitous and, in the often intolerable heat of high summer, it has felt as if I have been acting out "Thomas Hardy meets Lawrence of Arabia", with carthorses instead of camels.

Pigs know how to cope with the heat. Alice, our sow, gets her snout under the water trough, tips it over and then directs her triumphant nose towards the sodden ground. There she digs a hole, and wallows. Had the heat continued I would have joined her.

Much sweat has been created by the apparently simple business of moving livestock from place to place. You may recall my lurid accounts of sheep-catching marathons and of desperate struggles to contain wandering heifers. Well, thanks to a reader's letter, I have a new approach. It involves taking deep breaths *before* attempting to move stock, especially pigs. The result is less panting afterwards.

The rules have been laid down by a Mrs Mainstone and I feel I should pass them on.

Rule 1 of moving animals, she says, is: never use visitors to help. By and large, I agree. We were lucky in that a soothing and courteous art dealer happened to be passing by when the piglets were last shifted. We cannot expect a person of such natural sensitivity to be around every time.

Rule 2: Use only *one* local boy or girl, aged eight upwards, whom you know to be obedient, eager, purposeful and calm. I applaud the advice, but following it has been a problem. Children of the generation who got their stimulation from watching *The Flowerpot Men* may have been fine, but the modern *Star Wars* kid only wants to "zap" things. And zapped piglets fly in all directions.

Having got the herd on the move, Mrs Mainstone recommends walking ahead of the sow, banging a galvanised bucket of feed with a stick and crying "Tig, tig, tig." Infallible, she claims.

However, I suspect she succeeds not by obeying any of the aforementioned rules but by religiously observing her third piece of advice: "Explain to all concerned that you have in your mind a calm picture of success."

This works, and it has changed my life. I have taught a horse to walk, unguided, between rows of growing crops by never allowing it to cross my mind that he would not be able to do it. I have even loaded my three wild heifers into the lorry on my own — there was never any question that they would go anywhere else other than up the ramp. It is a sort of hypnotism.

The 'fluence works on turnips, too. Depressed by the heat, they succumbed to mildew. Conventional farming would have dosed them with fungicide. I merely filled my mind with a "calm picture of success" and did nothing. They withered even more. But then came a shower of rain. Now they are thriving.

And so is the succulent kale. One farmer near here told me that his entire crop had died. He asked what fertiliser I had used and what my spraying programme had been. I told him that I had done nothing at all, except kill the weeds with a horse-drawn hoe.

I was on the point of suggesting he tried a little hypnotism, filling his mind with thoughts of success. But I bit my tongue. It sounded dangerously like recommending that we talk to the flowers . . . and we know only too well the trouble that can get a man into.

CHAPTER FOUR

Sales, Shows and Stories

Autumn 1990

I have just joined a sort of Gentleman's Club. It has none of the classical façades of its counterparts down Pall Mall: indeed it has no permanent premises at all — it wanders from field to field. We have no rule book or committee, but we have an unwritten code of behaviour. If any member should break it, he would find himself beyond the pale. We are the Farm Sale Followers: we go from one farm auction to the next. Occasionally we buy something, but really the camaraderie is the thing, not the commerce.

Farm sales usually take place in September, at the end of the farming year. New owners are moving in and an auction is often the only way the previous owners can have a clear-out. It is a time of mixed emotions: often the farm is being sold as a result of a death, a retirement or, more generally these days, a bankruptcy. However, we rise above it and, like mourners at a country funeral, we can be boisterous in praising what *has* been.

I'm sure I shall not be in breach of our code of conduct if I outline how a farm sale proceeds. It starts at 10.30 a.m. sharp, so we gather at the refreshment caravan about half an hour beforehand. The man who fries the sausages could tell a story or two; in the course of his greasy business he is the first to know which farmers, whose harvests were not up to expectations, will be ordering the cheap burger rather than the costly jumbo hotdog. As a newcomer, I order a slice of his unrevealing fruit cake.

Then we parade around the field where the lots have been laid out. Whether it be a glossy combine harvester or a bag of old bolts, each lot gets individual attention. And individual derision. Like a blast of icy wind, the breath starts to whistle past teeth: "Phew, 'av yer se'n that there tracta? Hell, I shan't bid him more'n five quid for that." We all laugh, helplessly. Fellow members expect it.

The lots are usually arranged so as to start with the junk and climax with the seductive machinery. Beauty is in the eye of the beholder, though, so I find the junk more interesting. This is how you find a discarded horse-drawn plough or set of harrows, or other treasures. My eye was drawn the other week to a device described in the catalogue as a "horse-drawn strawberry hoover". It looked like a pair of embracing iron bedsteads.

Lot 1 is always the tangled heap of rusty metal and it is often bought by a shifty figure who drives the scrap lorry but who, you suspect, is the owner of a Palladian bungalow in Essex. No trouble with Lot 2, either, which tends to be a heap of old wood. The ever-optimistic auctioneer will say: "Fine load of timber. Who'll start me? Fifty quid. Come on. Fifty quid?" There is never a response. The shifty figure who bought the scrap mutters: "Two quid." The auctioneer quickly shifts his ground. "All right. Two pounds fifty. Who'll give me two pounds fifty?" And so, penny by painful penny we rise as high as four pounds.

But the shifty man always gets it. What does he do with it all? Then you get the buzz. Something inside whispers, "You've got to have this lot, whatever the price." At the last farm sale I went to it was a pair of hurricane lamps. I saw myself on freezing winter nights, with the power lines down, working my way through snowdrifts to rescue stranded sheep. I imagined the great swinging shadows of our horses as I tramped through the stable with a goodnight bale of hay. I had to have those lamps. The heart started to pound. "Pair of lamps. Twenty quid?" the auctioneer offered. "A quid," shouted a bright spark. I shot my arm into the air, not trusting any of that flickering eyelid stuff they go in for at Sotheby's. I offered another pound. So did the other chap. Battle raged. By the time we were up to three pounds he could no longer stand the heat, and I bought them for four. "Good pair o' old lamps, them are," a bystander said. Club members always congratulate a successful bidder.

I went on to buy a potato riddle, a horse-drawn hoe, a set of hurdles for containing sheep and two old tin cans. When it was all over, I walked down the field to pick up my lamps. They were gone. Nicked, I supposed. It was crushing to think we might have a member who would stoop so low.

Later that night the phone rang. A chum had picked up the lamps for me; thought they looked a bit unsafe. Happy once more, I eased into the leather armchair and thought to myself: "It does pay, to belong to a good club."

There are days when I wonder if this farming experiment of ours is worthwhile. The work is now getting harder, with fields demanding to be ploughed, stubbles to be cultivated and corn to be sown. I am beginning to get a truer taste of what it is like to work a farm by carthorse: it has been mere boy's work in the lazy days of the summer, but now the land needs full-grown men and I wonder if I am up to it.

I have become obsessive about the lack of rain since my farming career started seven months ago. Our land is so hard that, with horses, I cannot plough it deep enough or break the clods into a fine tilth; and the days are ticking away and soon it will be too late to plant anything at all.

Animals need to be watched carefully now. When they were on the summer meadows they fed and fended for themselves and an occasional glance was all the stockmanship that was required. Now they are in yards, bedded with straw, with every meal brought to them like demanding guests in a country hotel. Any delay and they complain as loudly.

After a hard day's ploughing, followed by a trip up the hill with the heavy pig-slop buckets and half an hour of mucking out, it is tempting to head for home. But it must never happen. The night that you succumb and think, "Oh, those sheep will be all right for tonight", you will pay the price the next morning. They will have skipped through their electric fence in search of an early breakfast and given you an hour's chase before the day's work can even begin.

So, at the end of the day, I sometimes wallow for a short while in self-pity, ease my boots, think of mustard baths, and wonder why I am bothering.

All I need, though, is some small confirmation that our experiment in turning back the farming clock is not entirely wasted. Fortunately, there are plenty of nuggets of news that suggest I am on the right lines. Much of what I read that masquerades as "new research" is nothing more than what I am re-creating out of the past.

I read in a farming newspaper, under the headline "It pays to be kind to livestock", that far better returns are to be had from pigs which are treated kindly. It went on to report that pigs thrive if given space to roam. Well, it may be news to them but it did little to stir Alice, our Large Black sow. She lives the way in which pigs were intended to live. She and her piglets wander on grass, hoover up the occasional acorn or windfall apple, and generally enjoy life. Alice is now in-pig for the second time, having reared all eleven piglets from her first litter with no veterinary attention of any kind, no needless jabs, no fuss, no ailing. I call that kindness rewarded.

The other day I met an old stockman and told him of this new piece of research. I have never heard a man laugh so loud. The report went on to say that milk yields vary by up to 20 per cent depending on which stockman handles the cattle. He chortled: "I could 'ave told 'em that too!" If the newspaper ever reports research into the sucking of eggs by grandmothers I promised to tell him.

Meanwhile, he informed me that any sickly sow should be turned on to fresh grass and given a shovelful of coal dust to root in. "Best tonic, for pigs," he said. I expect to hear the coal dust theory from research scientists any year now.

I have also been amused to read that some "serious" farmers have decided that, by using older varieties of wheat, they would need to add less artificial fertiliser and so save money. Farmers long dead were probably laughed at for sticking to those old varieties when the new ones first arrived. Pity they are not with us so that we can apologise.

I also read with interest of the organic experiments by the Duchy of Cornwall at the Highgrove farm. The people there told us that, "Good old farmyard manure is the foundation of their policy", and that it is their practice to "cultivate the leys and allow them to rot under a layer of muck before ploughing in September and planting corn". Such a practice is described in my battered, musty copy of *Stephen's Book of the Farm*. It was written in 1824.

So, all in all, despite a week which has left me with more than a few moments of despair, I sense the tide turning. I may be alone in walking the furrow, but I am not alone in spirit. I think we will hang on in here after all. It would be unkind to Alice to give up now.

There was an old ploughman who lived in these parts who was known to all as "Spittin' Mayhew". He was by all accounts a charming fellow, a great ploughman, and — despite his name — not a man of particularly unclean habits.

He won his nickname because of his peculiar way of starting his horses on their trudge down the furrow. Instead of getting hold of the rein and calling "G'up", or "Grrr", or some other throaty growl, he would simply spit lightly into the palms of his hands before clutching the wooden handles of his plough. Over the years, his horses got to know that the smack of their master's lips meant work was imminent. No other command had to be given. It must have looked like magic to anyone who did not know him; like starting a car by merely jangling the keys.

Carthorses get to understand all manner of obscure sounds and signals, but horsemen do have to learn a rather odd language. For example, to get horses to go to the right, we in Suffolk make a noise which sounds like steam escaping from a boiler: if it had to be spelt out it would probably look like "Wheeesh". If you are turning horses at the end of the furrow and wish them to come round without moving forward, you ask them to "Uh back" while at the same time persuading them to do a bit of "Wheeesh". If the ploughing is hard and you and the horses happen to be panting a little, the whole operation can sound like a vintage steam rally. Mr Mayhew's spitting system was far more efficient.

His name came to my mind this week after a tragic incident. We have managed to sow successfully five acres of vetches, a succulent green plant that can feed sheep, make hay, smother weeds, and fertilise the soil: a handy sort of crop. After the sowing came the rolling, to press the precious seed into the soil. To achieve this I attached a heavy ribbed roller behind a hitch-cart. This is a modern device which enables machinery normally drawn by tractors to be pulled by carthorses, providing it is not too heavy. It has only one drawback. The seat is made of two planks of the hardest of wood, and driving it over a cloddy field is as comfortable as taking a church pew on the Monte Carlo rally. In anticipation of a long and bumpy ride, I took a bite of lunch with me, wrapped loosely in a plastic bag.

If only the wrapping had been more secure, tragedy would have been avoided. Intent on watching the horses and making sure that every square inch of field was rolled, I failed to notice that every clod of earth over which we rode bounced my precious lunch ever closer to the rear of the hitch-cart until, having scaled a particularly nasty lump of clay, the rapid downward descent of the wheel finally propelled my bag of victuals in front of the heavy roller.

I saw it fall and called "Whoooh", but it was too late. A fine pork sandwich lay limply in the earth, the stuffing rolled right out of it. My generously thick slices now lay as thin as Parma ham. The apple was well on its way to cider.

As I rolled hungrily on, I bitterly reflected that it would never have happened to Spittin' Mayhew. His great distinction, other than that which gave him his name, was that he was probably the only ploughman regularly to carry a handbag. He had learnt what I am just beginning to discover: that when you set off to a field with a pair of carthorses, the last thing you want to do is return home before the job is finished. After all, the round trip could be several miles. Hence the handbag, which had in it all that a man might need to get him out of any tricky situation that fieldwork could throw up.

I am now drawing up a list of contents for mine. Ideally, I should find out exactly what Mayhew had in his, so I am scouring the memories of those who knew him. One day, when ploughing, someone lost a button off their trousers. "Now hold yer on a minut'," Mayhew said, and produced from his treasures a needle and thread. On another occasion, some machinery snapped, and out of the famous bag came a brace-and-bit. But it is easy to get carried away. I am tempted to stick by the old horseman's rule that all you could ever need was a shilling, a shut-knife and a piece of string: string to mend the harness, the knife to cut it and a shilling for a pint of ale.

When the drought finally broke and the rains came it was

a great shock. At least, it was to the sheep. During those arid days of summer they developed a cavalier attitude to electric fencing. Now they are having to think again.

In the dry weather, despite my test-meter showing that the wires were carrying a vigorous 5,000 volts, the fence meant nothing to them. It was perplexing at first to see flashes of blue electric spark where the wire happened to catch a twig or a branch, and yet see the sheep not only wander through the same wire but actually stand with it across their backs, unmoved.

At one stage I thought we might have the basis of a circus act: "Roll Up! Roll Up! See the Amazing Electric Sheep. Untouched by any Voltage Man can Create!"

The electric pulse comes from a mains-driven unit near the kitchen which gives a resounding tick each time it fires a few kilovolts down the line, which is about every second. It is like living with a grandfather clock. The one day it stopped, during a power cut, the house seemed eerily silent. After the third sheep escape, I checked to see that it was operating at the correct voltage. It was, and this was confirmed by the pigs, which are also contained by electric wires. They have only to make the lightest of contact between damp snout and cable for them to leap in the air and quickly reassess all thoughts of wandering. But the sheep kept plunging through: their fleeces, and the ground, were simply too dry to conduct electricity.

Then came the refreshing rain. Blades of grass sprang from the brown crusts that had formed over the meadows, limp green leaves on the mangel-wurzels awoke with new vigour, the kale stood to attention and even the depressed turnips started to show signs of life, despite my having made a mental note to hold a memorial service for those particular victims of the drought.

And then the sheep started to leap as well. The laws of physics had reasserted themselves and what went up was beginning, at last, to come down. In our case, 5,000 volts were going up the wire and coming down to earth via the freshly damped fleece and feet of the sheep. Having made accidental contact with this wire myself, I leapt in sympathy. The sheep have not escaped since.

Everything displayed a new vigour when the drought broke. Having stood idle for weeks, the carthorses have been able to harrow and roll the land to get it ready for some late-season planting of fodder crops. Iron-hard clods of earth which would not have yielded to the blow of a sledgehammer are now melting into crumbs under the influence of an inch of rain. A week ago, farmers round here had almost given up hope of getting seedbeds fine enough to sow winter corn. Now they are saying they have never had better. As in politics, a week is a long time in farming.

But it is the piglets for whom I am most pleased. Having been deprived for most of their short lives of one of their vital natural functions, they are at last able to root. Until now the ground has been too hard: soft black noses made little impression on the concrete-hard soil. I remember how, when they were only a few weeks old and still in the sty with Alice, their mother, she taught them to root in the straw. They would follow her example of bending the head and shuffling the nose forward. If any of the youngsters failed to join the lesson, Alice would waddle over and press their heads to the ground until they got the idea: pigs have a national curriculum, too.

Sadly, however, I thought the pigs might reach maturity without ever having put those hard lessons to any use. But when the rain came, the snouts were down, ripping up the grass in the orchard, tearing up the juicy tap-roots of the docks and generally making a joyous mess.

"You'll be happy now we've had some rain," said my wife, who has suffered my moans about the weather for months now. "Yes," I replied, "but we don't want too much till we've finished planting the corn." The dishcloth flew across the kitchen with the speed of a damp sheep fleeing the electric fence.

I was in a state of nervous agitation, my stomach gripped by waves of fear at the thought of what the morrow might bring. It was the most severe case of first-night nerves I had ever suffered, and it is all due to a long-standing ambition to achieve no more than a straight furrow, ploughed with horses, and be judged on it.

I had entered the Great All-England Ploughing Match. It seemed a good idea to enter in those far-off months of summer when the application form slipped through the door. Later the thought haunted me. Here on the farm I know that I can plough a perfectly decent furrow, but before a critical audience of several thousands I found the prospect unsettling. Forget any idea of rural peace and philosophy; there is as much competition in horse-ploughing as in any sporting event. The eye of the horseman as he glances down the furrow is no less keen than that of the snooker player lining up his cue: the ploughman's hands on the reins are as sensitive as those of any racing-car driver on the steering wheel. He is partly a mechanic as he adjusts and tunes his plough, partly a telepath as he communicates silently with his horses. Do not, whatever you do, attempt to speak to him.

The first time I tried to draw a straight furrow the communication was somewhat lacking. I called to the horses to "Gee up", but they did not move. With a little more urging they leaned forward into their leather collars, but it was clear they were of the opinion that the novice voice behind hardly deserved support. When we eventually got under way, I was so intent on watching a white stick that I had placed at the other end of the field to act as a marker that I failed to notice that my plough was maladjusted to the point where it was sliding over, rather than into, the earth. As I reached the far end, I turned and saw only two sets of hoof prints and a slight dent where the plough had ridden insolently over the soil. I expect I shall dream about that tonight.

The most frustrating part of ploughing is always that moment when, brimming with hope and anticipation, you turn back to look at the furrow and find it has a drunken roll rather than a Roman straightness. If I ever write an account of my ploughing days I shall call it "Look Back in Anger".

However, it is as much the horses as the men that draw straight furrows. I was told early on in my horse-drawn farming career that "a good carthoss needs no guiding," and you have only to watch a pair of horses at plough to appreciate the truth of that. Once a straight line has been scratched along the ground, the furrow-horse will follow it with an uncanny accuracy. Not only that, if he is a good horse he will place one foot in front of the other as he walks, so as not to spread his hoof marks on to the unploughed land. He knows when he has arrived at the end of the furrow and, almost without bidding, will start to turn. The horse's obedience to the line, however, can be the ploughman's downfall: all furrows emanate from the first one. It is like laying kitchen tiles: if the first row is not straight the rest will not be.

At ploughing matches it is not always the showiest horses that win, and that pleases me. Throughout the year, proud owners take their highly groomed prize mares and geldings to agricultural shows to compete for supreme championships. But when it comes to ploughing, all the medals and gloss in the world count for nothing if the horse cannot put one foot accurately in front of the other. So tomorrow I shall be sharing the field with all manner of carthorses which would be laughed out of the ring if they ever presumed to aspire to the Royal Show. But this will be their day. They may not have the best of harness, or the glossiest coats, but they can walk in a straight line, concentratedly and with goodwill.

The night before, I went through my list of packing which includes a set of heavy spanners, spare coulters, a couple of mighty nuts and bolts and a large sack of energy-giving oats. I tried hard to remember all the names of the parts of the plough so that if I was asked, "How far yer hake's snotched over?" or told, "The share's a bit proud of that sod", I should be able to hold my own. Trying to be a jolly ploughman is a serious business.

I slept badly.

We piled the horses and plough into our lorry on the Saturday afternoon and it was falling dark as our shaky entourage made a rattling furrow through the outskirts of Guildford, in Surrey. It was pitch black by the time we discovered the farm which had offered to put us up for the night. The farmer was a horse-ploughman as well.

Star and Blue, our pair of Suffolk Punches, were liberated into a grassy meadow for the night. We went inside the house for a hearty shepherd's pie and baked beans. In the midst of it, like the ghost of our own six-hour journey, the throb of an elderly diesel engine was heard advancing up the drive. It had come from Wales, bringing Joe. He had a face as weather-worn as a Welsh mountain, and as few teeth as a broken-mouthed old crone ewe. He had been going to ploughing matches for years but this, he said, would be his last. He was reminded that he said the same thing last year. Once liberally lined with pie and beans, he struck out again into the dark to look for beer. He refused a bed and said he would be happy sleeping on straw in the back of his lorry.

I was given the luxury of the guest room, but might as well have been on straw. All night, in the manic half-state between waking and sleep, I adjusted my plough a thousand times, harnessed imaginary horses, had nightmares about crooked furrows. My insides twisted with a mixture of fear and anticipation.

The crowing of the cockerel and the melody of Welsh voices dragged me out of the nightmare at 6.30 a.m. We arrived at the showground to find fifty pairs of plough horses expected to compete, and a crowd of 5,000 forecast. When the steward announced ploughing could commence there was no mad dash: old hands at this ploughing business finished their tea, dropped the collars over the horses' heads and strolled out to the pitch with the confidence that experience brings.

I had entered, with six others, in a class for those who

had not won a prize in any previous match. We emerged with the wasteful energy of youth and proceeded to attack the ground with our ploughs. The old boys at the far end were still leaning on theirs.

I won't go into details, although I know that sports reporters should, because describing your ploughing is as gripping to outsiders as people remembering their chess moves. Suffice it to say that my first furrow was straighter than I had hoped, the next one did not do it justice, and it was a good three furrows before I felt that Star, Blue and I were showing our best form.

Then I took a break and spied on the old hands. Heavens! There were some cracking good furrows; some straighter than a draughtsman could draw, and turned over with an evenness that left the ground looking like newly woven corduroy. Joe from Wales was ploughing like an old trouper; singing rather than commanding his horses but stopping now and again as if racked by stomach-ache. Was it the effects of the beer from the night before?

Some ploughmen tuned their machines as carefully as if they were musical instruments, never going more than a few yards without adjusting something or other. Often it would be a mere quarter turn of a nut, sometimes as drastic as a wheel moved by a whole inch.

Back at my end of the field, where the novices were ploughing, we worked on the principle that if all was going well we left it alone.

Slowly the untouched field turned into a ploughed one as each ploughman's work moved, furrow by furrow, to meet his neighbour's. It is the final furrow which is the most critical, for a bad finish to a bit of ploughing is as unsatisfying as a weak chapter at the end of a novel. I commanded the horses forward, closed my eyes and prayed. I did not glance back until the final furrow was cut. It was a good one.

The judges did a lot of pacing, a great deal of measuring and plenty of hard looking. Then the winners were announced. We came second in the novice class — the boys were magic, as they say — and we were well pleased. Joe did not win a cup, but the announcer congratulated him warmly on "ploughin' so well despite his hernia".

I have never believed in palmistry, but having spent a week confronted by wavy lines as intricate as those in the palm on any hand, I am beginning to change my view on the subject.

The whole thing started with my mangel-wurzels. I sowed the seed in May, with two horses harnessed to a ridger, a machine that looks like a plough but throws the soil out on both sides, leaving long, narrow mounds of earth. You sow the seed along the top of the ridge in the hope that the loose, crumbly soil beneath will give the roots room to spread. The theory is that the mangels then grow huge and fleshy until October, when they are harvested.

For a change, on our farm, the theory worked well in practice.

Intensively nursing my mangel-wurzels has given me six months of happiness. I have strolled along the rows kicking the tops off invasive thistles, pulling impudent docks and cursing the rabbits which find the sweetness of the mangel a tempting service area for a quick bite as they bound past on their way to destroy some other crop — we have an old sandpit which hordes of pestilential bunnies have come to regard as their home.

Only one other thing has marred what has been a successful crop. Just beneath the oak tree, halfway along the field, the neat and orderly rows of mangels display a violent swerve to the left. For their entire length they sit as neatly as cats' eyes down a motorway, but for some reason, just as they reach the foot of that tree, the plants show all the symptoms of having stumbled across a contraflow system. Without warning, the rows jump six inches to the left, then to the right, and then resume their orderly travel for the rest of the length of the field.

Ever since the seed first sprouted, I have had to live with this glitch. I have tried to fathom why it happened. Did the horses leap sideways, spooked by the ghost of an old ploughman? Was I looking over my shoulder in the hope that a cup of tea might be heading in my direction?

I have also had to put up with a certain amount of leg-pulling from the older farmhands who stumble down our lane. "We allus liked to get a bend in them there rows," they say with a smirk on their faces, "'cos we allus reckoned them blasted rabbits would trip up and break their necks." Then they roar with ancient laughter.

In fact, history records that it was always the habit of farm-workers, on a Sunday, to walk along the lanes in their best clothes spying on their neighbours' work and criticising the straightness of furrow and sowing. No wave or wobble went without teasing in the pub that night.

I have been doing a little snooping myself lately, and it is highly revealing. As you would expect, the farmers who are stable individuals inevitably sow their seed with military precision. Pythagoras would be proud of them: no geometrician would draw a straighter line along the ground than these men with their seed drills.

Then there are the others: men whose minds are clearly in turmoil. It may be their marriages, or their bank balances, but it shows.

If I had a caravan on Brighton pier and invited farmers to show me their fields rather than their palms, I would be able to tell you the ones who will be facing an uncertain future. They scribble the seed into the ground with the abandon of a young child discovering pencils. I would tell them to go away and reconsider their lives.

All this gave quite an edge to the sowing of rye, which we did the week before last. With the nervousness of a self-taught mystic turning the first of the tarot cards, I hooked the horses to the seed drill and called, "Gee up". I had convinced myself that the pattern the seed drill left behind would be a portent of my farming future. Either we would have a year in which everything proceeds on the straight and narrow, or we would face severe ups and downs, if not retreats.

Now, after a warm spell and heavy rain, the seed has sprouted; the lines of destiny have spread out across our little bit of landscape.

After careful study, I predict that I shall shortly be meeting a short, dark stranger. I expect him to be elderly, and he will stand at the end of the field viewing my lines, and smirking. Rabbits should expect stumbling times ahead.

If you have never read the farming press, perhaps I can save you the trouble. They are journals of unremitting intensification: "If you're not getting More, you're getting nowhere" might be their motto.

This week I have read how to increase birthrates in pigs by 0.2 per litter; been told that for only £5,000 I can buy a machine to suck forgotten grains from the bottom of storage bins. The big news from Denmark is that they have invented a "Fat O' Meater" to judge when pigs are ready for the butcher.

Well, my old sow tells me she has no use for 0.2 of a piglet; the chickens, at a capital cost of £2.75, see to it that any spilt grains are hoovered up, and when an old and trusted stockman turns up and declares in his broad Suffolk accent that "Them hogs, them be right ready", then I shall ring the butcher. The farming press and I live in different worlds. Their journals are designed to seduce and titillate with promises of More corn by using More chemical spread by bigger machines over More land. Soft cornography, I call it. I was on the verge of cancelling my order but then I would have no reason to visit the newsagent on Fridays, and who would collect the *Beano*? The young-stock would object.

But I have gleaned one good idea from the farming magazines and I must thank them for that. It came in an article written straight from the broken heart of a family farmer who had suffered poverty and hardship to pursue his dream of organic farming. His plight was desperate but his solution was inspired.

He was struggling, like I am, to farm on his own; he can't afford full-time help and neither can I. So, he invented a fantasy farm-worker called "Reg". Reg is ever-willing, always available and there is no limit to his skill. If you get up in the morning to find the cows have broken the fence, you think to yourself, "I'll get Reg to fix that," and continue your breakfast. The imaginary Reg is a helpful man to have around.

Now this fictional farmhand is working for me too. I have a corner of the yard littered with bits of old bale-string and paper sacking for Reg to tidy up. There is a catch on the stable door held by only one screw — I thought I might get Reg to fix that as well. Reg has broad shoulders. I swear it was him and not me who left the chain off the gate the other night and let the horses roam. He got a good telling-off but was still bright and cheerful when I asked him to creep into the pig shelters with fresh straw. Funny, it's one of those jobs I never got round to, but Reg seems to thrive on such tasks.

This week, my dream of a full-time helper came true. Reg ceased to be fantasy and became fact. He was in the form of an eighteen-year-old schoolgirl who wrote asking if she could give us a hand. I jumped at the offer. She arrived with an armful of A level geography textbooks and a keen desire to be around working horses. Without doubt, she knocked Reg into second place. After only one demonstration of harnessing, she was able to get horses ready for ploughing and invaluable when moving sheep from an exhausted pasture to a fresh field — not an easy job on your own, even with a good sheepdog like our Flash. She eventually established a good working relationship with the piglets who, I fear, are at a difficult age where their ever-growing strength combined with an acute lack of table manners can easily have you head over muddy heels if you are not careful at feeding time.

By the end of the week I was sad to see her go but gladdened by a chance remark she made as we drove to the railway station. "Your animals", she said, "aren't like animals I've met on farms before. They're not aggressive, they're calmer. Content — that's the word." This was music to my ears. More uplifting than any of Reg's imagined grumbles.

Then she made another fundamental observation, which showed she had really understood what we are trying to do on this little farm. "All your animals", she said, "do other things apart from just being fed and sent to the butcher." She's right. Our sheep are currently tidying up and manuring a stubble prior to ploughing. When the pigs have finished work in the orchard, I have a piece of land which they will make ready for next year's potatoes. As for the contented cows, they are making rich, steaming farmyard manure for which the whole farm will be grateful.

Reg is back on duty now. I don't expect he will come up with any such profound observations. I've decided that he can make the weekly trip for the farming papers. I suspect they are both of similar mind. Neither of them see any point in what we are trying to do.

I placed an advertisement in the local paper this week, short on words but awash with contentious undercurrents. "Wanted: Farmyard Manure (not poultry) 50 tons." My advertising campaign had a desperation to it, for without a substantial input of what is known in the trade as FYM this traditional and organic farming enterprise of ours will grind to a halt. I *had* to find some muck somewhere, somehow — my soil is crying out for it.

Muck is not very fashionable at the moment. Big arable farmers no longer bother with muck-heaps. A thriving and steaming dung-pile used to be the hallmark of the diligent and caring man; but round here such stinking blots on the landscape now mark you down as an old-fashioned sort of chap. But to a farmer like me who believes in returning life and nutrition to the soil, a muck-heap is a glory and a dream. Ambitious farmers like to boast of their trout lakes, grouse moors or their Four-Wheel Drive. These matter little to me. "Show me your muck-heap!" I say, and I want a true measure of the man's worth.

By my own exacting standards I don't rate very highly, but we are at an early stage in our farming career. A good muck-heap, like a pot-belly, sits better on an older and wiser man. Good muck takes time to produce, and loving care too. It needs a lot of straw which has been well trodden and dunged by cattle or horses for an entire winter. In the spring it should be dug out of the yards, turned, and allowed to compost before being spread. We haven't been farming that long — hence the advert.

I have to confess that I didn't hold out great hopes of getting what I wanted. Farmyard manure is quite rare these days simply because there aren't any farmyards. There are concrete pads on which sit humming grain silos and intimidating hunks of machinery — but they're not farmyards. You don't find chickens pecking at haystacks on such farms, only peacocks strutting around the lawn. Depressing places.

To produce what I consider to be proper FYM, you must keep your stock in the traditional way; but modern pigs are kept in housing where the floor is made of slats, beneath which lies a dank and malodorous lagoon of liquid and solid waste. The resulting noxious cocktail is called slurry, and I didn't want any of that.

But even if some cattle-yard muck was on offer, there could be problems for a truly organic grower. The farmer might well have wormed his cattle with wormers that would kill the dung worms as well. These worms are the organic farmer's best friends for they turn muck into compost. We want them alive and squirming.

Poultry manure presents a moral dilemma. Organic farming's regulating body, the Soil Association, has banned the use of it because they believe that by relieving intensive poultry farmers of their muck, we are helping to support enterprises of which organic farmers should disapprove. I disagree. If I were an intensively reared turkey, I would face the Christmas oven with happy resignation if I thought my short and tedious life had at least added fertility to a few hungry acres. However, the Soil Association makes the rules and we must stick to them. Having grappled with all the issues surrounding my innocent advert, I waited for the replies.

Sure enough, my phone was alive with calls from desperate men. I refused enough slurry to refloat the *Titanic*. A lady rang and said she kept horses near Stansted airport and was worried at the growing height of her muck-heap. I would have liked to help but 150 miles is a long way to cart muck, even to prevent a most unpleasant landing.

My quest ended with a phone call from the neighbouring village where lives an old horse-breeder who has carelessly tipped stable muck for as long as anyone can remember. Now he wants rid of it, for free. It covers the best part of a couple of acres, is five feet deep, and has rotted to the consistency of rich chocolate cake. It steams when forked like a freshly boiled Christmas pudding. It is *real* farmyard manure, all I ever dreamed of. We are about to start the long, hard slog of bringing it home. I am more thrilled with it than if it were a thousand acres of grouse moor or a few miles of the Test. Each to his own.

Author's note: Very little of what happens in the outside world permeates the thick red brick walls of the farmyard but it was inevitable that the momentous events following the downfall of Mrs Thatcher would have some influence on farmyard life.

The only item in our old stable that relates in any way to the latter half of this century is a transistor radio. I switch it on for the early morning news and more often than not it stays switched on. However, I have decided that I must be more careful about leaving it blaring. The political upheavals of the last weeks have proved infectious, and the farm is now in the midst of its own leadership battle. To predict the outcome, as ever, you need to know the contenders for the crown and the deviousness of the electoral system.

When we had only two horses life was simple. Punch was premier. Punch is a good-looking horse, intelligent but self-willed. For a decade he has been paired with Star as his deputy: as good a workhorse as anyone could wish for. He has never been known to be bad-tempered, or ever refused to pull with all his might. I heard that a previous owner had accidentally driven Star into a ditch so deep that it needed the fire-brigade to drag him out. When the rescue team arrived they found the peaceful Star up to his knees in mud, eyes half closed, blissfully enjoying the fresh grass growing up the sides of the deep drain and blaming nobody. It took a crane to lift him out, after which the old statesman was put back between the shafts and ambled off as if nothing had happened. It was his finest hour. When he dies we may have to have him sculpted in bronze.

But as in politics, so in the farmyard: it is not always the best man that wins. For most of his working life the down-trodden Star has been ruled by his chippy companion Punch. Punch can be bad-tempered, and, sensing when it is time for work, he will fling his head to the rafters to make it more difficult to get the collar over his head. When ploughing, he will irritatingly stop half-way along the furrow if *he* thinks it is time *he* had a rest. You may curse, even scream, at him but he shows his mastery of you by insolently moving off as slowly as he knows how.

Back at the stable he ruthlessly reaffirms his status by pawing the concrete floor with his front feet if he is not fed before the others. He knows that, to a farmer who has to pay ever-increasing blacksmith's bills, nothing grates like the sound of needless scraping of a £10 iron horseshoe.

So the old ruler reigned unchallenged till the spring when our new young horse arrived: Blue. I have already detailed the fierce and bitter battle that raged the night he moved in. Equine teeth were bared and those who feel wounded by political back-biting can think themselves lucky they didn't have the angry Punch coming at their spines with mouth wide open. He won the first round. There was no second ballot.

Things settled down for the long hot summer and each horse made his own territory on the meadow, Blue standing apart from the others, younger and fitter, with his flowing mane. It looked as though nothing could upset the old order. But when the season turned and the days grew shorter, I brought the horses back to the yard for the winter. Punch, at fifteen, is looking old now. It is rare for him to go ploughing, for Blue and the ever-youthful Star make an easily worked and efficient team.

Blue, sensing that Punch may be losing his grip, is seizing his opportunity. Hence the leadership crisis. We have a hayrack at which Punch and Star used to feed with Blue only approaching when they had finished. Now I notice he is standing his ground, and if he gets there first he will not budge. It means more back-biting, more aggravation. Fearing an accident, I have been putting a separate pile of hay in the corner away from the rest, so that he can feed in peace. But as soon as the horses are released from the stable Punch, I notice, walks straight over to his rival's hay and piddles all over it. His political techniques would be the envy of even the most ruthless Chief Whip.

I am not quite certain which of them will emerge as the new leader. When I open the stable door in the morning, I closely observe the order in which they file through. Blue, I'm sorry to say, is still last. Surprisingly, Star is occasionally first. Perhaps the dark horse will make it after all.

Of course, it matters little which horse decides it is boss of the herd for the farm is in reality a dictatorship. If any of them start getting above themselves I warn them that the phone number of the cat-meat man is never far from my mind. However, dictators have been having a difficult time of it lately and with the anniversaries of events in Eastern Europe due to be celebrated, I think the radio must be silenced. It gives the carthorses too many ideas.

Can pigs swim? There is a general belief that they can't,

because in the act of paddling the sharpness of their front trotters would slit their throats. This may be bogus folklore or true, but if the rain does not stop soon I may be able to answer from experience.

The torrential soakings of recent days have turned the field where the pigs live into a black, slimy swamp. You can't walk through it any more, you can only paddle. If you could make soup out of coal, it would look like this once verdant patch of land. Given that our pigs are black to start with, and that my only pair of wellington boots happens to be black, you can imagine what feeding time is like, particularly when it is getting dark. I have found myself kicking my left foot with my right, thinking it to be a greedy pig, while at the same time a confused and aggressive snout has been nudging me in the ankles in case my boot turns out to have a competing appetite. As this glutinous ballet is being enacted, I live in fear of being eaten in a frenzy of porcine gluttony. I've had enough: I've rung the butcher.

But I should not be telling you any of this: I'm afraid I'm coming to the conclusion that most people would rather not know how their food came to be produced and would prefer to erect a Chinese hedge between what goes sizzle in the pan and what went grunt on the meadow. Or even just grew: mushrooms, for example.

A couple of weeks ago I reported on the magnificent heap of rotting stable manure on a neighbouring farm. For six days, Gary carted 300 tons of it up here, and by the time he had dumped the last load you could hardly tell him apart from his precious cargo. In a distinctly hands-off managerial capacity I dropped in to see how things were going and spotted a cluster of mushrooms growing on the heap. I like mushrooms, and so do my family, usually.

I picked them and carried them home in my cap in anticipation of the sort of welcome that man the hunter might have had on his return to the cave. But it was not to be.

"Mmmm," my wife said, sniffily. "Are you sure they're mushrooms?"

I was confident.

Our boy, aged eight, took one look and asked: "Where did you get them, Dad?"

"Off the muck-heap," I replied.

"Well, I'm not eating them," he said. "They'll have germs."

"They're disgusting," our six-year-old daughter added.

I dabbed the mushrooms lightly with a damp cloth, remembering that they should never be washed, and fried them lightly with butter. It is a long time since my tastebuds had had such a treat. "Mmmm," my wife said, "very earthy."

The phone rang: it was London friends. "We're just eating mushrooms . . ." choked my wife, ". . . he found them on the muck-heap."

"Oh no!" shrieked the distant voice. "Will you be all right?"

I, too, choked. In outrage. "Where do the mushrooms in your smart London eateries come from?" I called across the room. "Grow in those little blue boxes on supermarket shelves, do they?"

I fear that most people now believe that food is born and bred in packets, and anything that is not vacuum-packed is second rate. My mother used to make exquisite Yorkshire puddings from flour, eggs and milk, until ready-mixed ingredients appeared in the shops. Now she will use only those: they're packaged, so they must be better. They're not.

All this is bad news for farmers like me, who trudge valiantly through mud to feed our pigs on natural barley, or cattle on oats and kale. We pursue the production of wholesome, unpolluted food with a religious fervour, only to find that customers can't stomach the real thing. It is like the case of vegetarians who insist on organically grown food. Do they know that organic growing depends on what comes out of the back end of animals which are reared to be killed in the prime of their lives? I think not.

But I have delicious plans for the pigs. I have delved into aged tomes and discovered recipes for the curing of hams and bacon. I am planning a smoke house where flitches will hang and absorb the subtle aromas of smouldering chips of oak.

I mentioned this to a butcher. "Yes," he said, unmoved, "you can do it that way — but we've got this chemical you can just paint on. It gives it the colour and it gives the bacon flavour." What sort of flavour? I asked. "Oh," he replied, "a taste just like the real supermarket stuff . . . . That's what people like."

Eventually we killed our first two pigs, which was distressing, but not for the reasons you might imagine. What started as a brush with the nineteenth century ended up as a head-on crash with the confused values of the late twentieth.

I always ask for neighbourly help when pigs have to be marshalled: capturing agile swine calls for a man of dogged determination, instinctive stockmanship and an ability to curb his tongue in front of the children. I have none of these qualities, and Richard, my neighbour, has. By chance we also had a sculptor staying with us who claimed to have wide experience of pig-handling, having spent time in the peasant cultures of mid-France. I was more interested in the muscle which years of chiselling had bestowed on him. Then I eyed the two, long, lean hogs, and set up hurdles ready to catch them. A bowl of barley meal was the bait.

For a long time I have wondered how I would feel when the first stock that I had raised would be heading for slaughter. After all, these were pigs from Alice's precious first litter. I had been with her on that sunny June morning when she had effortlessly delivered them into the world. We had cared for them like babies, thought of starting a photograph album of their piglethood. We loved those wriggling youngsters: they were the first star attraction of our farmyard.

To my surprise, I felt no remorse at their going. I can put my hand on my heart and declare that no pigs have had more comfortable, cosseted or better-fed lives than these. As the only purpose of raising pigs is for them to be eaten, I faced the abattoir with a clear conscience, with one provision: they must die as they had lived, with dignity.

For the moment, however, they were still free. They edged towards the bowl of meal but sensed the hurdles were some kind of threat, and the slightest twitch by any one of us made them flee. Even a six-month-old pig is unstoppable if it has made up its mind to be free. The sculptor advised, the neighbour acted and I let them get on with it. By macabre coincidence, the travelling pork butcher arrived with a wicker basket over his arm to inspire us with his hams, chops and sausages. The killing of the pig was always a great occasion in the small farmer's economy, with the whole family turning out to help: what had been a grunting, well-fed friend by the back door would shortly become their guarantee of food throughout the winter. I have read lurid accounts of the slaughter, of weeping children holding jugs to catch the blood, of bladders being excised and used to hold the lard. I was happy to delegate all these tasks to the abattoir.

While the pigs' attention was diverted, we seized our chance and snapped shut the hurdles. We were half-way there. Recalling his Gallic adventures, the sculptor suggested we put their heads in the bucket of meal and, by applying gentle pressure, back them into the trailer. Minutes later we were bound for the butcher, five miles away. The slaughter house lies hidden behind the white-washed façade of a Suffolk village. The beasts are killed by the son of the vicar, who spares time to advise and sympathise with first-timers like me. More importantly, animals get kind attention, too. There is no stressful overcrowding in undersized pens: animals are killed within a couple of hours of being delivered. It is as far removed from an insensitive factory atmosphere as you could wish to get. So why, and here the twentieth century intrudes, will the institution probably be forced to close? It is to do with 1992, when the whole of Europe will break into blandness. The rules that apply in Naples will be the same as those in Norwich; abattoirs that kill 100 animals a week will be wrapped in the same bureaucracy as those that kill 1,000. But rules made to govern a steelworks would never work if applied to a blacksmith's forge, and neither will the rules of mass meat production ever allow small men, like my butcher, to survive.

The argument is long, but I know of no more sympathetic or stress-free end to a couple of pigs' lives than the one enacted last Thursday. Is there any chance that somewhere in the vastness of a united Europe there will be room for the concerned farmer who wants a civilised end to his animals' lives? Is there the remotest possibility that the small and caring may ever be valued as highly as the mighty and efficient? Pigs might fly.

I gather from reading the women's pages lately that it is fashionable for men to confess infidelity and let their emotions hang out. A stiff upper lip, they tell me, is the worst thing I can adopt. So, if you will afford me a shoulder on which to weep, may I confess to having had a brief fling?

She was no beauty. Indeed, in any farming set-up other than ours she might well have been written off as scrap. The blue enamel of her youth had been severely scarred and her general appearance suggested a tired old barmaid who had, in her time, been backed and bumped into just about everything. Dented but unbowed, she stood foursquare in the barn; a wise old biddy who had comforted many a pressured farmer through difficult times. With a mechanical wink and a nod, she was now offering the same solace to me. On her radiator, she called herself "Fordson Major". Mrs Robinson would have been more appropriate: she was a mature temptress of a tractor.

I dislike tractors, always have, and have said often that if I ever had to give up using carthorses, farming would no longer hold any attraction for me. But last Monday morning, with the icy rain thrashing in off the North Sea, I looked at the deep, sodden muck in the yard, glanced at the tractor and its hydraulic muck-fork and decided that, for a morning at least, I would make the starting of the diesel engine my first job of the day.

The bliss of the union did not last long. I felt uneasy at filling the farmyard with vile exhaust fumes. It was as crass as blowing cigarette smoke into a florist's shop: acrid fumes are no fair exchange for the pleasant odours of equine and bovine flatulence, blended with the sweet smell of hay.

Then there was the noise. The rattle of the old tractor echoed between the high walls of the barn and obliterated the comforting sounds of animal teeth ripping hay from a bale, and the odd metallic chime of a carthorse's hoof as he moved in his stall.

Besides, I soon began to do serious damage. The tractor has a high frame to stop the driver being crushed should it topple over. Good idea, except when you reverse without looking and take a length of guttering with you. In my haste to retrieve that situation I drove forward with such speed that I was unable to stop before the sharp prongs of the muck-fork had turned the drinking trough into a colander. Enough was enough.

Ten minutes later the affair was over, the tractor was back in the barn and the horse was harnessed to the tipping muck-cart; I was back in the bosom of my old love, happily wielding a fork beside a big, warm, patient horse. There was no grinding of gears or revving of engines, just honest labour and gentle understanding.

A good carthorse can be worked far more easily than

any tractor. As I progressed across the yard with my muck-shifting I did not even have to lead him forward. If you say, "Juss w'un step, ol' hoss," he will drag the cart forward a single pace, and stop. "W'un more, old man," and he will edge forward again. He listens as you curse when wet muck falls from the fork and splatters your face with filth, sympathises when a heavy forkful sets you panting. There is no such friendship in a tractor.

Just to check that my marriage to horse-drawn farming was still sound, I went to the Royal Smithfield Show at Earls Court later in the week. To lovers of the latest in high-tech farming this is a veritable bordello of temptation, with machines in iridescent livery all promising hard-pressed farmers to deliver more for less effort. It is seduction on a grand scale. I found it a thoroughly miserable experience. Only when you come close to these monsters do you realise how far farming is now removed from the grasp of the common countryman. There are huge devices for spreading animal feed which must require so much attention that the animals hardly get a second glance.

Tractor cabs are now so high and insulated that it seems unlikely that young farmers can ever develop the same understanding of the soil as their ancestors, who walked the furrow behind a horse-drawn plough. And, for all the majesty of these modern machines, no salesman standing beside one ever seems as proud, to me, as a horseman standing at the head of his plough team.

To fall in love with devices that tempt you into believing they will ease the burden of working the land is very easy. Having strayed down that path, I know now that happy farmers are those who keep both feet on the ground.

* * *

I hope the vicar didn't notice, but during the carol service the other night I was shifting as nervously as a troubled schoolboy. It was simply that the carols struck home in a way they never had in the days before I became a farmer. In nearly every hymn, a verse or a phrase set me off on an anxious train of thought, each one leading back to the farmyard.

No sooner had the boy soprano cut the air with his "Once in royal . . ." than we arrived at the lowly cattle shed. As the proud owner of several cattle sheds of the most lowly state imaginable, I did not find any comfort in being reminded of the work, and money, needed to keep them standing until next Christmas. As for "Where a mother laid her baby/In a manger for a bed . . .", it suddenly came to me that it was in the woodworm-ridden old cattle manger that I had left the spare breast for the plough. I've been looking for that for a week. Ah, the plough! It was a tradition in these eastern parts that good ploughmen should have turned all their land by Christmas Day, and to bring them luck for the following year, they would sleep on Christmas Eve with the breast of the plough beneath their beds. The congregation had reached "And our eyes at last shall see him . . ." by the time I had been through all the parcels of land that were still unploughed. If I'm in bed with the plough by Easter I shall be lucky. Feeling weak at the thought, I was glad when the vicar asked us to be seated.

But there was no rest. One lesson later "In the Bleak Midwinter" were announced, and the organist attacked the opening notes with an enthusiasm that suggested *he* hadn't been forking twenty loads of horse-muck that morning.

". . . Earth stood hard as iron, water like a stone." I shuddered — a mighty dread had filled my troubled mind. In a big freeze-up, which is bound to come some time, gallons of drinking water will have to be carried bucket by bucket from house to farmyard. It comes to us via an electric pump in a well and if the power lines are down I shall be the one who has to tell the thirsty horses, cows, pigs and piglets that water is like a stone. Peace and goodwill will soon evaporate in an unseasonal display of foot-stamping and snorting.

"We three kings . . ." intoned the vicar. Not much better. To be truthful, I've had a bellyful of wise men out of the east. I suspect there is a roving pack of retired farmworkers who hunt me down, not to dispense wisdom, but merely to haunt and undermine me. They stand watching me plough, and ignore the straight and neat furrows, but remark when one of them is less than perfect. "My ol' dad, he'd say that look like a dog's piddle in the snow . . . " they declare, and burst into a laugh so deep that you know it is coming from the heart. Then you plough a near perfect furrow, but they won't say anything about that one.

Our next hymn was "It Came Upon a Midnight Clear". My ranging mind swung towards Alice, our sow, who is due her second litter on New Year's Eve. I know she will have them at midnight because it was at that time she started when her first litter was born. Except that was in June, when a midnight dash to check the sty was quite pleasant. If things are going to get deep and crisp and even, Alice may have to improvise. I have already intimated that there is ample precedent for mothers having to manage when there is no crib for a bed and I think Alice has got the message — I caught her shunting straw into corners yesterday.

"The cattle are lowing . . ." sang the choir. Of course they are. The bull arrived last week and his presence has put my maiden heifers completely off their food. Where once they used to issue a coarse, rasping "mooo" towards feeding time, they now moan a seductive melody which leads me to believe there will be little trouble from them this Christmas.

Heartened by that thought, the Christmas spirit briefly wafted over me. Even when the soloist rose to sing the "Boar's Head Carol" I didn't allow the thought of a freezerful of unsold pork joints to disturb me.

Then we sang "While shepherds watched . . ." and I remembered the battery on the electric fence. It is flat and the flock will be roaming. God rest ye merry gentlemen? Some hope.

CHAPTER FIVE

Winter Days

New Year 1991

This is my first New Year as a farmer. The last nine months feel like a decâde: hardly a minute has passed during which some obscure agricultural point has not been occupying my mind. But what seemed at the time to be moments of indecisive agony fade into insignificance when compared with current anguish. There's always one problem that seems bigger than the last.

Currently, the sheep's dental arrangements are giving me cause for concern. Having sown turnips in the spring, hoed them through the long hot summer and watched them alternately flourish and perish as each mean shower of rain gave way to searing heat, I now find them ready to feed the sheep. I have arranged matters on a cafeteria basis, simply turning the sheep on to the field and telling them to get on with it. Since they had never seen a field of turnips in their short lives, it was like watching children trying to master the removing of a top from a boiled egg.

First of all they trotted around the field, bleating. Then they looked at me, longing for some instant junk-food from a bucket. I gave them a stern lecture on how I had slaved all summer to fill their winter bellies with succulent turnips; they stared blankly. Like a mother who had baked fresh bread only to be asked for Mother's Pride, I stormed off, warning them that they had to eat what they were given or starve. The warning was sufficiently sharp for even Flash the sheepdog to look upon the flock with sympathy.

As I have discovered, with farm animals it is all or nothing. Once the ewes had got the taste for the turnips, the entire field was cleared as if some giant vacuum sweeper had gone over it. Except that, on close examination, I can see that the turnips have been gnawed level with the ground but no further. Half of each turnip still lies buried, uneaten. I tried my matronly approach and ordered them to dig up and eat up their greens. It had no effect.

Now, the question is: will the sheep, when hungry, dig down any further to remove the other half of the turnip or do I have to go round picking them out one by one, turning the running buffet into a laborious silver service? I don't know the answer.

Of course, if it rains it will make the roots looser in the ground, which would be good. But if the land gets too wet, the sheep will ruin it. So I had better hope for a freeze-up. Except that then the turnips would be rooted for ever and I would have to play waiter again.

I'm coming to the conclusion that farming is one long conundrum which is never ever solved. So I don't take decisions; I gamble. When I choose badly I can always claim in mitigation that it would have worked fine had it not been for the wet/dry/thundery/hot/cold weather.

If this seems like a cowardly way out, I am now of the belief that this is the way farmers have been operating for years. A clue came in a poem sent to me by a Suffolk farmer whose mother originally wrote it. As I didn't get round to sending any Christmas cards this year (too busy shouting at sheep) may I offer you these verses both as a memorial to my first twelve months as a farmer and a dour hint of things to come:

> The Farmer will never be happy again,
> He carries his heart in his boots.
> For either the rain is destroying his grain
> Or the drought is destroying his roots.
>
> He will tell you the spring was a scandalous thing
> For the frost and the cold were that bad.
> While what with the heat and the state of the wheat
> The summer was nearly as bad.
>
> The autumn, of course, is a permanent source
> Of sorrows as black as your hat.
> And as for the winter, I don't know a printer
> Would print his opinion of that.
>
> In fact when you meet this unfortunate man,
> The conclusion is only too plain.
> That Nature is just an elaborate plan
> To annoy him again and again.

Happy New Year.

I am starting the new year on an unlucky note, I think. Matters that should have received detailed attention over the festive period have been allowed to slip, and I fear the worst for the coming farming year.

You see, Father Christmas brought me a book of Christmas superstitions in which I stumbled across a reference to the yule log. I have several chunks of tree which no man has been able to split asunder, and any of these logs would have been a prime candidate for the job of yule. Apparently, had I slung it on the fire and allowed it to smoulder for the twelve days of Christmas, saved the ash and sprinkled it with the seed when the corn is sown in the spring, I would have been assured of a bumper crop. I didn't, and now I am worried. Neither did I weave my drunken way round the meadows sprinkling spiced ale, so no doubt we shall have a further year of lacklustre grass to look forward to.

Farming superstitions must be treated with the greatest respect, I believe. Having discovered, while ploughing, several flints with circular holes in them (called hagstones in East Anglia), I could not bring myself to do other than hang them above the horses in the stable. It was believed that witches came in the night to ride the horses unless the hagstone was there to provide protection. Strangely, we do have a horse which has been found sweating in the morning, as if having been ridden. The vet found nothing wrong with him. It has not happened since the hagstone has been swinging over his stall.

High-technology agriculturalists will, I know, be laughing like a drain at all this. They farm not by the portents, but by their dull "protein analyses" or their tedious "dry matter content" and all the other jargon. But even the high priests of advanced agriculture are beginning to admit there may be forces which are beyond them.

This week, I read that much of the data on which organic farmers base their choice of crops may be suspect. If you are a conventional farmer, you can choose the seed you want by referring to extensive and reliable tests. Organic farmers look at the same research, but for varieties of seed that resist diseases. This is essential since we cannot use sprays to kill bugs. The tests have been on seeds grown conventionally, however, so when they are used in an organic system they may behave in a completely different way.

I must admit it is easy to be seduced by the velvet voices of modern farming. Even seeds of such dull crops as turnips and cattle beet are sold with all the panache of a box of milk chocolates. I have just read a brochure designed to tempt me away from the mangel-wurzel. This is a vital and traditional crop on our farm: it grows easily all summer, is stored in the autumn and is then fed to sheep, cattle and carthorses in the dead days of February and March. It is, to them, like a bite of fresh apple amid a diet of dry muesli. Now I am being offered a "super-new" and "improved" variety. The description, I am sure, was written by the same gentlemen who dream up claims for detergent.

Well, I'm not giving in. Mangel-wurzels have been the staple winter fodder in these parts for more than a century, and this accumulated wisdom and experience has got to be worth something. It may not be that you can analyse it, but we dismiss the old approach at our peril.

If I have a new year resolution, it is to make ever greater strides backwards. I am doubtful whether futuristic farming has anything to offer that will ever bring farmers to a closer understanding of the mysterious processes of growth. Organic farmers have fought a long battle to dispel an image of themselves as masters of "muck and magic". I can't imagine why. I find farming to be an endless series of magic tricks played out on a well-dunged stage.

On which subject, may I beg a round of applause for Alice, the Large Black sow, who last week produced a litter of twelve? It is considered lucky for something black to be first across the threshold in the new year, and I did think of inviting her in . . . but I decided there might be just a little too much muck mixed with the magic.

I could swear that the days are getting longer. I am sure that a couple of days ago I was able to cart feed to the sheep at half past four when the week before I wouldn't have been able to find my way.

In some ways, the dark depths of midwinter have been a disappointment. Perhaps I had read too many Victorian farming tales: I expected my winter routine to begin with grooming carthorses before dawn, and be ploughing before the sun is over the hill. I have read accounts of ploughmen seeing their way by the light of the sparks rising from the feet of the horses as iron shoe struck flint. Of course, if you're a lazy son of the late twentieth century you stay in bed and miss all the pyrotechnics, as I have done. I am resolving to get up earlier.

"Won't matter what time yer' get up," an old farmer said to me the other day, "we don't have no darkness in the country. Not any more." He went on: "It's these blasted bright lights they have outside these 'ere pubs. Them throw their lights for miles. Why, folks 'ave even got 'em on houses. Blasted things they are." I swallowed heavily, and said nothing. I had put up two last week.

But I can see what he means. Unlike the old-fashioned pearl bulb, these quartz-halogen lamps are unsubtle in the way they fling their rays of light. The farm next door even has an automatic one which switches itself on when it detects movement. The trouble is that it cannot differentiate between burglars and hedgehogs. Out for an innocent midnight stroll, Miss Tiggywinkle suddenly finds herself spot-lit as if for the big solo in *Phantom of the Opera*. There is no comfort in these piercingly bright lights: they suggest hostility, whereas a single glowing filament of the old kind offers comfort. A weary traveller looking for shelter would surely be drawn to a warm, yellow glimmer rather than to a shrieking, electric-blue stab. My new lights may yet come down.

But just to satisfy my curiosity, and to check on whether darkness does exist in the country, I walked to the top of the farm after dark but before the moon had risen. It was a clear, star-lit night.

As the old farmer had predicted, the countryside was ablaze with more of these blasted floodlights than there were stars in the constellation of Orion. Their light travelled for miles. I don't mind the ones that illuminate tricky road junctions, but why should I have to endure the glare of a light four miles away which serves only to illuminate somebody's Ford Escort as if it were a work of art? By the time all this wasted light has been added to the intrusive glare of the Sizewell B building site, which already fills our eastern sky, darkness seems a thing of the past.

Does it matter? Strangely, German farmers seem to think it does. They have made the remarkable discovery that weed seeds that have lain buried cannot germinate unless they come to the surface and perceive instant light. Consequently, if you stir up the top of a ploughed field in the dark, the weed seeds which you bring to the surface will not sprout. If you are farming organically, without any weed-killing chemicals, this is an important step forward — providing that the steps forward can be seen without stumbling.

Very often, much of what is hailed as a new discovery in farming is a resurrection of an old and forgotten practice; but I can find no reference to this one in my library of venerable farming textbooks. The only story I have come across is of a farmer who was so delighted with a newly invented plough that he took his horses out at midnight and ploughed by the light of the moon.

So, I am planning midnight sorties when the time comes to harrow the land ready to plant the corn seed. How we shall find our way around the farm is far from certain: the horses may have to be equipped with miners' helmets. (No, put away the notepaper, it wouldn't set off the weeds. The harrows would be dragged behind them.) But for the experiment to work we need darkness in the countryside again. Surprisingly the power-station builders have already co-operated by dimming their lights a little so as not to confuse migrating birds. I am not quite sure how to put my case to the owners of all those patio floodlights. Still less to the restless, stage-struck hedgehogs who set them off.

February 1991

A few days before the Gulf war broke out a note came through the door to tell us that, because of the gravity of the situation, the church would be open for prayer and its floodlights (kept for special occasions) would remain switched on. At the moment, our village has no vicar and so the gesture comes from deep within the hearts of the worried parishioners, many of whom remember the war of fifty years ago.

The old farmworkers who stroll along our lane and pause for a chat when they see me at plough have spoken often of war in recent months. The commemorative Battle of Britain fly-past in the summer lined up more or less directly over our farm for the flight into London. Scores came out to watch. I had a pair of carthorses pulling harrows that day, and I remember an old boy turning to me and saying: "I was 'arrowing during Battle of Britain, I remember. But them planes, they wus flying t'other way then." We laughed a little and then silently watched as the unfamiliar drone of propellor-driven aeroplanes came nearer. The horses were uneasy: they are used to the scream of the modern jet-powered American tank-busters that often come and play directly over us. But they sensed something about this fly-past was different. Like them, it was out of its time.

Now the old boys are thinking of war again. Current conflicts they cannot grasp, and so old memories are flooding back. Of course, in the Second World War, farmworkers were heroes. A booklet came my way this week which is the official history of British farming between 1939 and 1945, called *Land at War*. It begins: "Though the sword has long been a symbol of war, the ploughshare one of peace, today this symbolism is no longer true. For in the hands of the modern British farmer the ploughshare has become another weapon of war."

The ploughshare proved effective: Britain never starved. But it was deadly, too, and the price paid for victory on the land was huge. By being encouraged to plough up land which had never been under cultivation, farmers were given a taste of mass-production which was to turn into an addiction.

No doubt it was right at the time, but to those with a

modern eye it seems grotesque to boast, as this booklet does, that "the South Downs, unfarmed since Saxon times, are now under the plough", or "Feltwell Fen, a place of water, dykes and yellow reeds. Now the dykes have been cleared, the reeds replaced by acres of lusty wheat." The chapter headed "Reclaiming the Bad Lands" would make any 1991 conservationist feel faint.

Within thirty years, the farming heroes became the villains as calls for even more efficiency brought about the destruction of marsh, hedge, heath and woodland. One of my older visitors used to drive a powerful caterpillar tractor. "Smash anything up, that would," he told me. "I worked all round here. All them hedges that's gone, I pulled 'em all out. Now I suppose you'll be puttin' 'em all back. I can't tell you how many ponds I've bulldozed. Hundreds. Thousands of trees I pulled out."

This is a bewildering time. The old farmworkers see farmland that they fought to keep fertile now being encouraged to lie fallow under the European scheme to reduce the amount of food we produce. Not five miles away, good wheat-growing land is being planted with gorse to enhance a new golf-course. There are few animals in this arable area, but now we read that experts are advising farmers to consider mixed farming with a bit of corn, a few head of cattle, a flock of sheep and, perhaps, pigs — farming just as it used to be.

The last generation were taught that such old-fashioned practices could lead only to ruin. The new generation is being warned that the bandwagon that started rolling at the outbreak of war is finally coming to a halt. The old boys say farming is coming full circle, and we will surely arrive at the point from which they started when they went to war. "Why," they say, "there's even someone up the lane ploughing his land with hosses!"

And most of them would be overjoyed to think that the past might come again, were it not for the church remaining open for prayer, and the ominous floodlights blazing.

Even above the stench of a cloistered farmyard in mid-winter, I can still sniff an issue as it comes wafting down the valley, particularly one which carries the heady scent of showbiz about it.

Various perfumed Joannas and Glorias have signed a petition to support a ban on keeping pigs in narrow, hideous stalls, and the use of tethers to restrict what little movement the creatures have. The government has promised action, but not before the turn of the century; the National Farmers Union has adopted its traditional Luddite stance and opposed any change.

What none of the parties to this debate has done, however, is speak to the victims. I have.

Alice, our Large Black sow, is residing at what might be called her town address. Normally, she patrols the orchard, which she considers her rural retreat, but comes up to the metropolitan bustle of the farmyard when it is time to farrow. Her sty is not narrow and does not confine her in any way: it is roomier than many an Earls Court bed-sit.

In this cold weather she spends most of the day playing Garbo behind the door, emerging into the world only for sanitary purposes (always in the north-east corner) and for meals (taken in the south-west corner). Pigs are fastidious. Her outside run is bedded deep with soft oat-straw, so as not to be hard on her feet. She likes her comforts. While feeding her litter, Alice lies indoors in such a position that only the damp, slimy extremity of her black snout peeps round the door. It twitches and scans like early-warning radar. She can detect a swill bucket two counties away.

To get her view on close confinement I shut the door on her and bolted it. There was a dreadful shocked silence while she came to terms with the indignity. The delicate snout pressed against the door, but it did not budge. She grunted at it but it did not surrender. Then an eerie silence fell, broken moments later by the faint sound of what I swear was a whimper. It was not the squeal of anger she emits when moved against her will, nor the sharp reproving snort she fires off when more than ready for her meal. It was a whimper. A cry. No heart could remain unmoved. I opened the door. Interview over.

So good luck to the Joannas and Glorias and let all pigs be free, like Alice. I fear, however, that not everyone will rejoice when swine finally get their liberation. You see, on the sandy coastlands round here farmers find it better to keep pigs out of doors and free-ranging. They ring the fields with electric fencing and then scatter huts in which the pigs live: the huts are made of shiny corrugated tin, bent into a crescent shape with a simple wooden back and front. They look like small Nissen huts. Pigs love them but — and here's the problem — the neighbours loathe them.

The issue of these pig arcs "littering the countryside" has made one village near here kick up a stink more powerful than any pig could emit. "They are", bleat the humans, "turning our countryside into a silver city of pig arcs." I have heard it said by locals that some, if not most, of the complainants are "London-types".

Some public figures have been brave enough to advocate that confinement of pigs should continue. Edwina Currie appears to be persuaded by the argument of a farming voter who said that if we give up intensive pig production we will have to import more pork, which, since we would be the only country in Europe to introduce this ban, would mean that pigs elsewhere would suffer. I thought I might bounce this one off on Alice, but such a feeble justification for mass mistreatment of animals hardly seemed worth lugging across the farmyard, even in a swill bucket.

If ever there was a "pig-in-the-middle" situation this is one. Nice country-loving, green-wellied liberal thinkers are repelled by the captivity of an animal as intelligent as the pig, but equally revolted by the sight of them when they are let free. "It's not the pigs," they will say defensively, "it's the silver huts they live in. They're just too ghastly."

Well, having bought a pig arc recently, I must report that there was not a lot of choice. The Liberty-print pig shelter is some years away, and so somebody's going to have to lump it. I hope it is not the pig.

I mixed the pig-swill at nine in the morning and the tap water was running freely. By eleven the frosty, north-east wind had got up and reduced the flow to a trickle. By noon the pipes had frozen solid. That was a fortnight ago, and I am still carting six loads of water each day from home to farmyard in a ten-gallon tank slung between a wobbly pair of iron wheels.

I now have a routine. I clothe myself with more tweed

than normally worn by an entire shooting party, pass a hose through the kitchen window and fill my first tankful with warm water. Then I trudge through the snow with my precious, steaming cargo. As I approach the yard, sheep bleat with a cutting insistence, cows bellow their impatience, carthorses prance in anticipation of breakfast — which wakes the pig and her litter, who then rattle the door of the sty until it is almost off its hinges. I feel like the only waiter in a roadside café with a coachload of starving football fans to serve. And you should see the look on the animals' faces when I break the news to them that it is only water in the tank and that the food is coming later.

I cannot make up my mind about winter on farms. It is either a time of wild beauty to be enjoyed, or merely a period of intense drudgery. For rural writers it has always been the former, and they are invariably more eloquent in their descriptions of frosty, ice-bound mornings than they are in depicting sultry summer days. I can understand why. Winter distils all the elements of farm life into an intoxicating spirit, the vapours of which can be breathed as you crunch through the snow across a farmyard like ours. Senses are heightened: details too commonplace to be worth noting on ordinary days are observed and enjoyed when set against a background of intense bleakness.

For example, when it became clear that snow was imminent, I took a horse and cart to cut a few loads of kale so that whatever the weather there would be something fresh for the stock to chew over. The land was hard as rock, and as my billhook struck the earth when slicing through the stems of the kale it made sparks. The horse, with his tail to the bitter wind, dropped a heap of dung which steamed with such ferocity that I swear you could have boiled a kettle on it. On an ordinary working day I might not have noticed.

As we trundled home with the last load, the heaviness in the snow-laden sky became oppressive and I glanced at the sheep. My sheep are cunning and are well aware that they can manipulate their inexperienced master. It seems that it has only to cross my mind that they may need an extra feed for one of them to give a heart-rending, hungry bleat. On this occasion, I thought that our pregnant ewes would be better off in the shelter of the farmyard. The instant the thought occurred, the first flakes of snow fell and a cold, lone sheep ambled towards me with plea written across its face. Snow was being driven horizontally and the wind had an edge to it keener than any blade on a scythe. I set my sheepdog, Flash, to gather the flock and watched as he became a dark smudge moving swiftly through the white blur. Bringing sheep safely home in a blizzard is a memory I shall treasure.

So much for the romance. What about the harsh facts of daily life? Well, I do not feel moved to write in poetic terms about the early mornings when penetrating frosts bite into my cracked fingertips. I cannot begin to describe the gloom when I find snow has blown under the barn roof and covered several hundred pounds' worth of feed: it is no problem now but the feed could be ruined when the thaw comes. I will not describe to you the frustration when numb fingers are trying to undo a chain on a gate, or the feeling when the metal feed scoop freezes to your fingers.

And all this fumbling, shivering and lugging of heavy water buckets is set against a background of cows, sheep, horses and pigs all raising their voices and stamping their feet, demanding food to keep out the bitter cold. There have been times in the last few days when I have stood, frozen, in the middle of the yard and screamed at the lot of them. There was not much poetry in it.

I have scanned a wealth of wise words relating to management of farms, but nowhere have I seen published a fundamental rule which, this week, I broke. It goes: "No farmer should ever sit by the fire and think that all is well with his farm. It is asking for trouble."

Imagine the scene: the flames licking high, the sloe gin about to be broached, the snow thawing to torrential rain. I decided I would just feed the pigs before the weather got any worse.

The instant I got to the sty I sensed something was wrong. Those piglets know the sound of the kitchen door and even before I have got across the yard, they are squealing with sudden hunger. But this time they were silent. I assumed they were feeling the frost and were keeping their heads well tucked under mother's warm and ample belly. But when I poured the swill and still the frenzy failed to materialise, I got worried. Alice lumbered to her feet and took a few mouthfuls as if to be polite; but the litter of black piglets just remained heaped on each other in a pyramid, breathing deeply and looking like a wobbly blackberry jelly. Slowly they stirred and my worst fears were confirmed for those curly little tails which would normally be coiled as tight as bedsprings were hanging with a depressing limpness. As the rugby song has it, "When a pig's a failure it straightens out its tail. But all pig's tails are curly 'cos piggies never fail!" These pigs were sick. Failing.

The vet declared the fast-growing litter had a feeding

problem so I gave them all an extra dose of minerals and went back to the roaring fire. The hand reached for the sloe gin, and then I thought of the sheep. Better to feed them before dark.

Tightness gripped my stomach as I climbed into the sheep pen. We have a borrowed ram with the demeanour of a grumpy old major-general who has failed to grasp the idea of the vending machine and thinks you put your mouth under the spout. He does not understand that you eat the food from the trough and not from the bucket. In his eagerness he usually butts you, and it hurts. But this time there was no rushing, butting attack. He just stood by the wall, poor old soldier, trembling, weak and defeated. More feeding problems? The feeding of stock is one of the most intricate of farming conundrums that I have had to unravel. If you go to the merchant and simply buy a bag of ready formulated "sheep nuts" or "pig nuts" it becomes simpler, but I am reluctant to do so. It is our aim to feed our stock from what we grow on the farm. Economics force commercial farmers to do otherwise, with the result that what is fed to farm animals these days can be far from certain. So our stock eat oats, barley, kale, mangel-wurzels and sugar beet, which we do buy in, but it does mean that if the meals aren't carefully balanced you can omit some vital ingredient. Unlike humans, animals can't instinctively run to the fruit bowl or the larder when some internal alarm goes off.

I went to my book of sheep husbandry. I flicked through improbably named diseases from husk to hoose, blobwhirl to turnsick, and gid to hydatid of the brain. I decided he had all of them. I fell back in the chair, blobwhirled and giddy myself, and reached for the sloe gin. The phone rang.

"Pigeons are on your kale!" came the bad tidings. Kale, I might explain, grows a couple of feet high at this time of year and is the only green crop that stands proud of the snow. Hungry pigeons scan the horizon for it like snow-bound rail commuters seek the lights of a station buffet. By the time I had recorked the sloe gin and got up the field there were at least three hundred on a mere two acres. I called in the big guns. A neighbour in a farm cottage is a keen, if less than accurate, shot and has time on his hands. He built a hide out of fallen branches and ex-army camouflage netting, dug himself in with a flask of tea and waited for the next winged sortie. In they flew, beaks sharpened, to be met with a hail of gunshot most of which missed but certainly spoilt their appetites. The pigeons regrouped to plan their next attack. My man is on a constant state of alert.

And so am I. For although the piglets' tails have taken a distinct turn for the better and the old soldier of a ram is able to stand to attention once more, I am only too aware that an armed guard is required to keep the peace on this troubled little farm.

I was looking through a list of rare breeds: species that once formed the basis of farming in Britain and which are now so few as to be endangered. To my amazement and regret, I found it did not include the Farmer's Wife. If anything is rare, she is.

There are many women who think they are farmers' wives, but are not. If asked what they do, they reply: "Oh, I'm a teacher and I'm married to a farmer." Or, further up the scale: "I sit on the local bench and do a lot for charity. We farm, y'know." Neither is the sort of lady I have in mind.

Real, pedigree Farmers' Wives collect fresh eggs each morning from speckled hens, wear their hair in a bun, bake bread in white aprons, and feed pigs with a smile. They ease the farmer's boots from his weary feet and darn the holes in his socks with wool they have spun. They know every lamb and tend the sick ones; they clank across the farmyard with pails of milk fresh from Buttercup. And do they ever utter a cross word? No. They are too potent ever to need to lose their tempers. For, apart from being the farmer's domestic servant and support, their rural ancestors have bred into them formidable powers.

The Farmer's Wife will happily leave her baking to face an angry bull, or put aside her mending when her rough-hewn figure of a husband summons her to the barn to help drag a half-born calf from its mother.

Yet beneath this toughened exterior smoulder womanly passions which no son of the soil can ignore . . . I'm sorry. I've been reading too much Thomas Hardy.

However, the farming family is not what it was. Take my own situation, for example. I have a fine wife, and no intention of finding another. But she would be the first to admit she is no Farmer's Wife. Early attempts ended in humiliation as she produced from the oven a noble attempt at a loaf of bread. She thought it looked rather good, considering, but at that moment the neighbour (a Farmer's Wife) burst in and said: "Oh dear, you've 'ad a bakin' accident."

However, she possesses great patience. She does not complain at having to sit next to me in heated cinemas, even when the stench starts to rise from my manured boots.

She was in no way put out when I recently declined to escort her to a glittering literary dinner in favour of sitting with a farrowing sow. Such is her resignation that on a brief romantic trip to Venice she did not flinch when I pulled out my wallet in Harry's Bar and brought with it enough straw to deep-litter a cowshed.

She is even showing some positive promise as a helpmeet: after a spell during which the sheep escaped several times a day, she developed amazing reactions. One evening, I was phoning some friends and failed to get a reply. Casually I called: "I think they're out." In a flash she was through the door, waving a stick hysterically and shouting: "Where? Where are the woolly bastards now?" I doubt, anyway, whether I could do better on the open market. Indeed, I was shocked at the ad in *New Farmer and Grower* magazine in which some brazen woman wrote: "Well-conformed dam seeks pedigree sire of traditional breed to share good organic leys with a view to moving on to permanent pasture. Good performance testing essential. No bullocks."

Does this ad set the style of clinical courtship for the new farmer? Where is the stolen embrace behind the cart shed, the fleeting brush of lips behind the threshing machine, the plighting of troths in hedgerows? I know that modern safety legislation frowns upon a tumble in the hay, and that a furtive glimpse of a milkmaid is not the same now that she sits at a computer console rather than on a stool. Even so, can you imagine Jill Archer placing such an ad to woo Phil?

However, rare breeds have to be saved and I must act. In addition to our Suffolk Punches, Red Poll cows and Large Black pigs I intend to rear a proper Farmer's Wife. As I have no desire for a change, the current one is going to have to learn a few tricks. I shall start by getting her used to being called "woman", and having to ease the boots nightly from my feet. Gradually I shall work up to suggesting she gather wool from the sheep to darn my socks. I am not certain how she will react, but if I am not here next week you will know why.

CHAPTER SIX

New Lambs, New Laws

Spring 1991

I am thinking of spring lambs, of Easter, and how our little farm can now claim to be a greetings card come to life. The lambs started to arrive a week ago.

Although my calendar has had "Lambs from this date" inscribed on it for nearly five months, we were never convinced. We suspected the idle ram of having a permanent headache. He never looked remotely enthusiastic. But we owe him an apology: behind that indifferent stare burned a virile passion. His girls are now paying the price.

The problem with ewes which have not had lambs before is that they don't try very hard when it comes to the birth. They assume they are suffering indigestion, or perhaps constipation, and wander around rather stiffly putting a brave face on things. One of our ewes, half-way through giving birth, sprinted across to the trough for feeding time. The half-born lamb dangling from her at the time will now presumably be so traumatised that it lives in fear of the dinner gong.

When lambs are born without any bother, it is a delight to watch. Their confused little bodies plop onto the ground like a bundle of wet washing, and take just one breath before looking for Mother. Then they stretch their limbs and stagger to their feet, wobbling on all fours like a rickety table. One foot goes bravely in front of another and slowly they inch closer to the radiant warmth of the ewe. So begins the most vital expedition of their lives, to the teat.

Considering the ewes haven't had months thumbing through Mothercare catalogues and getting into the maternal frame of mind, they do get the idea remarkably quickly. The instant they have sniffed the lamb and confirmed it is theirs, they defend it with stamping feet if you approach too closely, and a headbutt will meet any other ewe that seems intent on theft. Once mother and lamb have bonded, I move them into the family quarters.

I have divided the lambing pen so that one half has ewes about to lamb, and the other has ewes with lambs. As soon as a new ewe and her lamb join the family circle all hell breaks loose. They start sniffing the newcomer, confused as to whether the lamb might be theirs. If they decide it is not, instead of leaving it alone they butt it into the air till the poor little thing retreats, bleating. This forces its real mother to step forward to the defence, providing she is not occupied in sniffing another ewe's lamb and thinking it might be hers. The great revolving mass of sheep and lambs, all sniffing and butting each other, gyrates for about five minutes until families are reunited.

But for all the lambs that are born without any difficulty, it is the ones that don't make it that stick in the memory. We lost twins because the first-born got its head firmly wedged and its legs tangled. It happened with another ewe but this time I was there to help. Remembering the lessons at our local agricultural college, I greased my hand. For those lambing lessons we were given an appliance made out of plywood and rubber which represented the rear end of a sheep. To give it an authentic feel, the rubber bag was encased in warm water at sheep temperature, and filled with slimy jelly. A bicycle pump was attached to it and the whole thing could be inflated so that when you inserted your hand as if helping a lamb to be born, the tutor could give a quick stroke of the pump to simulate a contraction. It is not the sort of device to which innocents should be exposed.

The real thing is far more satisfying if you are successful; and deeply depressing when you are not. I am just back from the lambing pen after helping a ewe to give birth to her over-sized lamb. The lamb didn't make it. I blew down its mouth as a sort of kiss of life, tickled its nostrils with a straw. It didn't respond. It was a big strapping sort of lamb that deserved to frisk in the meadows. Half an hour before I had been sitting in the bottom of a wagon, enjoying the lambs at play. Now I had a dead one to bury.

But if you can't take the knocks you are a poor shepherd. My nineteenth-century *Cyclopaedia of Agriculture* says of shepherding, "Owing to the nature of his work affording him time and opportunity for quiet thoughts, there has always been associated with shepherding much that is poetic and beautiful and good."

True. Except that this novice shepherd finds the poetic bits followed so swiftly by action that I am a spinning mass of emotions. Coupled with ewes and fresh-born lambs spinning around their pens, these are dizzy days on the farm.

* * *

In reporting on the birth of our spring lambs last week, I omitted to mention other harbingers of spring: namely Easter bunnies, spring chickens and daffodils.

Firstly, I am suspicious of the bunnies. Every night they try to dig burrows in the middle of the sprouting barley, but having scraped away the top inch of soil find the ground too hard to penetrate. By way of consolation they stuff themselves with my precious young corn. Just when I had seen off the winged black vandals from the rookery, nature has mounted a bob-tailed attack from the rear.

Not being anything of a shot, I haven't a clue what to do about rabbits. The birds could be scared away, but how do you frighten a rabbit? Read Beatrix Potter stories to it? I am told there is some kind of poison gas you can blow down the holes, but my inclinations are against using chemical sledgehammers to crack natural walnuts. One lad did offer to bring his ferret, but knowing my luck it would get lost, the lad would burst into tears and I would end up having to mount one of those headline-catching rescue operations which extract stuck potholers, while the rabbits sat there laughing.

I suppose I could always train a ferret myself, but I have little inclination towards rodents and even less to teaching them tricks. My aged copy of *Livestock on the Farm* devotes a whole chapter to Ferret Management, beginning, "One cannot picture any animal so relentless, so filled with dogged perseverence, bloodthirstiness, and sheer love of slaughter as the ferret." It goes on: "At feeding times, they should always be called with the same word, 'Puggy-Puggy-Puggy'." I discussed this with the family, who decided the "puggy, puggy" was dangerously close to our usual call of "pussy, pussy". Not a nice way for our cat to go.

So I decided against training ferrets and turned my mind to chicks. If I have had one humiliation in my farming career it has been my failure to produce a single egg for my own breakfast. I was given a bantam hen and clutch of chicks last year, and fed the brood up in eager anticipation of the day when puberty struck them and filled my frying pan each morning. Alas, every chick turned out to be a cockerel: noisy, thin, bony cockerels not fit for even the most poverty-stricken pot. My neighbour kindly agreed to despatch them as their 2.30 a.m. crowing was wearing him down as well as me.

He did a tidy job, but so impressed was the mother hen by the way in which her brood had been wiped out, that she upped and left and went to live with him. Women always seem to go for rotters with guns. I vowed to be thoroughly unpleasant to the next hen we had in the hope that she might take a shine to me.

Then, last autumn, I was offered a box of chicks all guaranteed to be hens. I put up a chicken-run made of tall poles and wire-netting and released them from the box in which they arrived. Like the start of a balloon race, they all took to the air and flew high into the branches of the oak trees, clucking contemptuously. I don't mind free-range hens; it is the free-range eggs that are a problem. The hens now lay mainly among the hay in the barn, so that whenever I pull down a forkful rotten eggs drop like stinking hail. We tried clipping the hens' wings so they couldn't fly. They flew.

Which brings me finally to the daffodils. So far, alone of all living things on the farm they are causing no trouble at all. From my window I can see them swaying. It might be a rabbit signalling all clear to his mates. Or there's a hen among them, laying an egg.

Some unsuspected scent in the air, or warmth in the wind, must have set an alarm bell ringing this week and galvanised me into a frenzy of farming activity.

In the spring, a farmer's fancy turns to thoughts of grass. To most town dwellers, grass has little romance about it, being merely a patch of annoying stuff which you have to mow. To farmers with livestock, however, a field of growing grass spells freedom, for when the grass is growing thick and fast, stock that has been confined all winter can be liberated to feast upon it and get fat. The season of bucket and bale carting will be over, the months of trudging through the mire of manure will be done, and the farmer will feel his holiday has begun.

The stock quite like it, too. So, with the warmth of the sun on my back for the first time this year, I harnessed Star and Blue, my Suffolk Punches, and went to harrow the meadows. Harrows are the farming equivalent of the garden rake: they are spiky, heavy, and as the carthorses drag them over the sward they rip out the dead growth of the previous season so that light and moisture can get down to the roots.

To transport the harrows up the lane without putting the horses into cart harness, I have built a sledge: an old door fixed on two log runners. Onlookers wondered aloud whether I now regarded the wheel as too new-fangled. I ignored them and got on with the job.

When harrows break the blades of grass a heady scent

rises, more intoxicating than the nose of a vintage wine. Tractor drivers would miss it, being enclosed in a cab and in front of the action, but as I plod behind my horses I am exposed to the full boozy waft of it.

I have learnt in my short farming career, however, that it never pays to get drunk on success. Farms have ways of sobering their masters. When I returned to the stockyard with the horses, out of the corner of my eye I glimpsed a ewe and knew she was not long for this world. Shepherds say that "sheep have only one ambition in life, and that is to die", and here was the proof. This poor animal, heavy in lamb, was swaying on her feet eyes glazed and staring, waiting for death. There was no point in calling the vet. I offered her feed but she showed no interest, and when she lay down it was clear she would never get up. So I was left with a corpse.

I remembered a discreet advertisement that used to appear in our local newspaper: "All classes of fallen stock removed." The ad was placed by the local hunt, which took away dead animals, such as my ewe, butchered them, cooked them and fed them to the hounds. You may find this distasteful, but as a method of disposal it had a lot going for it. There was no risk of infected animals littering the countryside, no danger of rotting corpses being buried near streams which provide drinking water, it also meant that valuable animals which could provide meat for humans were not being killed merely to feed dogs. The method was cheap, clean and relatively dignified. But no longer. I made a few calls and discovered that the dead hand of Brussels has pole-axed the animal disposal business. New rules, some unnecessarily tough, require hygiene standards far higher than most hunt kennels can afford or need; the same rules which are forcing small, and arguably more humane local abattoirs to disappear. However, where there's livestock there will be dead stock, and the disposal of "fallen" animals is becoming a big problem.

Dead stock were often processed for bonemeal and factories collected them, but fears about bovine spongiform encephalopathy (BSE) have brought that to an end. Hides and fleeces used to have some value, but not now because of cheap imports.

The problem is getting desperate. There are reports of dead sheep and cows being dumped by Scottish roadsides, thus shifting the disposal problem from the farmer to the council. What can the hill farmers do? They often farm on soils only inches deep over rock. They cannot afford to have the carcasses carted away.

One dead ewe is no problem for us. I dug a grave in a disused corner of the farm, harnessed a horse and loaded the sheep on to my sledge. In a sombre procession the three of us made our way across the fields. The joy of a morning on the meadows was gone; one harrowing experience had made way for another.

I shall never forget the day the American fighter planes returned to their Suffolk airbases from the Gulf war. In fact, I shall remember the precise moment: it was just as I opened my mouth to shout "Come by" to my sheepdog, Flash, that the deafening jets screamed overhead. My voice was drowned. The dog failed to hear the crucial command. Disaster ensued.

Flash and I were returning a ram we had borrowed. With

lambing time approaching, I judged that it was not in the ram's interest, or the ewes', for him to be present at the birth of his offspring. This may be the age of the New Man but I doubt whether we are yet ready for the New Ram. So we took him back to his flock.

Possibly, being penned up for weeks with young pregnant ewes chattering endless baby lamb talk had got too much for him. He certainly looked relieved as he turned his back on his flock of young wives and bounced happily into the trailer. But when he arrived back at his old home to find he was to be herded into yet another, even larger, flock of young, pregnant ewes, he winced.

Taking one look across the meadow at the flock, the ram decided any fate would be better than life with the girls. He bent his neck until his powerful head was in the butting position, turned his back on fatherhood, and made a desperate run at the flimsy electric fencing. Even had the fence been connected to the national grid it would not have stopped him.

I had to admire his determination. He was always a noble creature: black-headed and Roman-nosed, I had called him "Stormin' Norman" because, being of the Suffolk breed, he was very schwarz in the kopf. When it came to being first at the feeding trough he showed a similar spirit to his namesake in that he would go "over, under, round or through; whatever way it takes" to be first at the feast.

As he broke for freedom, Flash made a valiant effort to catch him. We almost had him cornered, and if the dog had obeyed my command to go to the left ("Come by") we might have captured him. But the dog couldn't hear me above the roar of the jets. Stormin' Norman was found the next day, invading a garden in the next village. He returned, somewhat sulkily, to barracks.

Back home, more trouble from black-headed creatures. Rooks by the hundred were gathering in the trees round the field of newly sown oats and barley, and when no human was in sight the cowards were swooping and devouring the precious seed. I could have wept; I spent five gruelling weeks ploughing that field with horses, walking wearily up and down it with harrows and seed drill. And now all that work was to be undone by the criminal beaks of the thieving rooks. I swore revenge.

A kindly neighbour shot a couple of them — who would have thought, a year ago, that I, a civilised man, would be so delightedly grateful for two dead birds on the doorstep? — and we pegged them out between stakes in the middle of the field; the limp corpses dangling as a reminder that rooks are not welcome. These marauders need to stare death in the face before they get the message. Even so, a few braved it.

Then someone mentioned a new bird-scaring kite, designed by a Second World War fighter pilot. I sensed that this particular war was about to turn in my favour. The black and yellow kite looks uncannily like a bird of prey; it hovers and shifts with the breeze and beats its threatening wings.

The unsuspecting rooks flew in from the north-west, heading for their regular feast. Then they spotted it, faltered, and, rather than land, circled. The bravest went close to the kite but lost their nerve and retreated.

I don't know what the rooks think this kite is, but it puts the fear of God into them. I shall no doubt lose support from the more sentimental wing of the conservation movement, but I do not care, as long as I can imagine these miserable birds fleeing back to their nests, terrorised to the point where their horrible, spindly legs shake at the knee and their nights are disturbed by visions of the flapping yellow and black-winged predator.

What matters is that my sky is free of rooks. The kite is keeping the peace. We have, as the real Stormin' Norman would put it, total air supremacy.

There was no celebratory cake, no popping of champagne corks or even a greetings card: my first year as a farmer passed unnoticed except by me, and it was only a passing thought. There is plenty of time to entertain fleeting notions as you harrow in seed with horses. It is a plodding business, silent except for the jingle of harness and the panting of carthorses; and not very accurate work either, for all we are trying to do is to drag the soil over the seed which has been put there in neat, ruler-straight rows by the seed drill. Harrowing is easy work which leaves the mind and senses to roam freely.

Today's harrowing is the final and crucial act in securing a crop of oats and barley, which we shall harvest in the autumn and use to feed our stock next winter. I want this farm to be sustainable and not rely on applied chemicals or bought-in animal feed. Twelve months of farming has not dimmed my conviction that this is the proper way to grow food. But the rooks that gather in the trees as I harrow have the power in their beaks to cause my system to founder, and I ponder this threat and how to meet it as I guide the horses up and down, up and down, up and down the field.

Of course, farmers used to be resigned to the theft of precious seed by marauding birds. There was an old rhyme:

> One for the rook, one for the crow,
> One to rot, and one to grow.

They may have nodded in polite agreement when their vicar sternly warned them that "whatsoever a man soweth, that shall he also reap"; but they knew from experience that it was only a one-in-four chance.

Children used to be kept home from school and forced to spend entire days running up and down fields banging sticks together to keep the birds on the move. With my mind now in a harrowing-induced state of free fall, I wonder if there might actually be some virtue in this. Surely city children would enjoy a spell of country air. And think of the exercise. Better than all that bouncing around on imported coconut matting. Environmentally friendly too, for the less seed that the birds nicked, the more the field would produce without the need for chemical assistance. Two more rounds with the harrows and my wandering mind has convinced itself that I shall write to the Minister of Education and suggest it.

Come to think of it, I could have asked the Minister of Agriculture to pass it on. We may be a small, misguided and antique operation but this did not prevent the Minister himself considering us worthy of a visit. He invited himself last Friday. We were quite flattered, but not as flattered as we would have been had John Gummer not also been our MP. He hadn't far to come.

I feel sorry for the man. Although he can boast the grand title of Minister of Agriculture, the fact of the matter is that if you consider the European Community as a supermarket Mr Gummer is no more than a branch manager. I get the impression not of a man in charge, but a rep sent out by head office. But that was not the most abiding impression the minister left with me. It took several more rounds of the harrows before my spinning mind eventually settled on what sets this man apart from anyone else I have seen on the farm this year.

It is his shoes. He has the shiniest black shoes I have ever seen. This is not something of which a Minister of Agriculture should be proud. Green wellies would have been marginally better, but not as good as scuffed boots with treads packed solid with manure and laces an inch thick where they have trailed in the mud. Mischievously, I invited him across the slimy yard to look at the pigs.

This was a good move from his political point of view for he could remind us that he had just announced a ban on the cruel tethering of sows. It was not such a good move from the polished shoes point of view. We approached the nine remaining recently weaned piglets. They miss Alice. From beneath their dangling ears, shiny black snouts sniffed eagerly at shiny black shoes. Then they turned away, disappointed.

No, I ponder, as I drive the horses and harrows round the field for the last time, nice chap — pity about the shoes. Start trusting a farming minister in shiny shoes and before we know where we are, we'll be confiding in a bank manager who wears trainers. Which rambling thought brought me to the end of my harrowing and my first year as a farmer. The field was as smooth as a freshly iced cake. I might put a candle in the middle of it and have a private celebration.

* * *

I went to the local cinema the other night along with an assortment of farmers, publicans and bed-and-breakfast ladies to witness a stunning performance. Whether it was black comedy or a horror epic we are still not certain, but not one of us was unmoved.

The star of the show was our young environmental health officer (something of a matinée idol) and his act was to explain the Food Safety Act 1990, which has just come into force. Why was I there? Because no one who handles food in any way can escape from its scrutiny. I wish to sell produce at the farm gate, but as far as I can gather even if you only squeeze out teabags for the WRVS, the new law is set to pounce.

The Food Safety Act is an indigestible feast of legislation, and in the room above the cinema it rapidly became clear that explaining it to us was a task which Cecil B. de Mille would have been stretched to cover. The show started with high drama as the EHO smote the blackboard with a rod and declared: "I am now empowered to remove from your premises any food which I consider suspect and place it before a Justice of the Peace at any hour with a view to having it declared unfit for human consumption." Best of luck, mate! If he thinks he's going to get a cheery welcome from the magistrates round here when he pounds the door in the middle of the night bearing a malodorous pork pie, he's got another think coming.

There was worse to follow. "Remember," his voice dropped menacingly, "from now on I shall consider water to be a food." We nodded, stunned. Warming to his theme he continued, "Listeria is everywhere. There have been instances," he took on his severest tone now, "where children have been allowed to lick bowls after mothers have been making cakes!" We reeled in horror. Then he showed us a diagram of a thermometer with the temperatures above which hot food should be kept and below which chilled food must be maintained. I noted gloomily that the temperature of the human body fell half-way between. I fear that any one of us may be placed before a Justice of the Peace at any hour of the day and condemned.

Then it was time for questions. We learnt that, yes, even village halls must get their kitchens up to scratch; we discovered that there is an exam we all have to take, to prove we understand what the new laws are all about. Quite frankly, I wouldn't wish to be the one who tells the lady who has made the teas for our local WI for the last twenty years that she's got to sit an exam on how to do it. I suggest the EHO gets his friendly Justices of the Peace to do that one for him.

Still, on balance, good luck to the Act. The old one, after all, had more loopholes than a rancid chunk of Emmental. It is no loss. And thanks to the EHO because for all its high drama it was a surprisingly good night out. We are now busy thinking how our little farm can comply with the new law, to enable us to sell our pork, lamb, sausages and potatoes. When we are ready I shall be more than happy for the EHO to come calling. I have nothing to hide. Except perhaps one thing.

Last week I bought an old bacon slicer, and I love it dearly. It had spent sixty years of its life in the village shop and then been put to rest in the retired grocer's garden shed. But when it emerged on Saturday its chrome still gleamed, its painted cast-ironwork was still a vibrant red: there was even an edge on the circular knife. I fell in love.

Now I have a vision. It is of the most delicious hams from our own pigs, smoked and cured on the farm, and waiting to yield to my new slicer, in succulent, moist leaves. But will the new laws allow it? The very best smokehouses, I am told, are built out of converted privies. As for curing, it means that chunks of pig must spend many long weeks slopping around in unsavoury liquids at whatever temperature God sends. Add to that the antique bacon slicer, and I fear my EHO may go into a faint. Never mind. When I'm hauled before the beak in the middle of the night, the bacon slicer's coming too.

Many years ago, in Sheffield, I tried to pursue a career in the meat trade, but it was short-lived and hardly a success. I was eleven and our butcher was short of a Saturday delivery boy. On being offered the job I swelled with pride until I matched the massive pork pies that adorned his window. I grasped the wicker delivery basket, and then my troubles began: I couldn't lift it. With a sigh, he unloaded lumps of ham and parcels of chops. Promising to make two trips of it, I set out.

The load was still too heavy. Stewing steak was soon rolling in the gutter and liver littering the pavements as the unwieldy basket swayed in my aching arms. In true schoolboy fashion I dropped each parcel at roughly the right house, rang the doorbell and ran. Once I got caught. Out of a scullery door charged a pit bull terrier of a woman: no teeth, hairnet and an odour of fried fish. "Ey up, tha's dr'pt 't," she screamed through clenched gums. I vowed from that day never to enter a profession that dealt with "customers".

But now I'm having to change my mind. The only way we can hope to make any money (or lose the least) on our small farm is by selling our meat at the "farm gate". We have meat of which we can be proud. We can guarantee crackling on our pork (remember that?) and flavour, because of the natural way in which our pigs are fed. We don't produce enough to offer it to supermarkets; not that we'd want to anyway, for I am sure it would be a huge success and then they'd want more of it and we would be headlong down the road that leads to Alice the sow and her daughters being moved from the shimmering cow-parsley in the orchard to some concrete establishment where they would never sniff the breeze again. So we think small: no twenty-first-century marketing man is going to advertise an international chain of butchers by boasting that it all began in *my* pigsty.

Thankfully, I have to know very little about butchery, except the words. The first time I took pigs to the slaughter the butcher asked: "How do want 'em doing?" He was clad all in white; more fitting for a wedding than the funeral of my beloved stock.

"Oh," I said casually, trying hard to exude the confidence of one who had been doing this for years, "just the usual killing and butchering, please."

"On or off the bone?"

"I'll have some on," I paused, "and some off."

He persevered. "Want the belly made into sausages?" he asked.

I was backing away to the car by now. "Ah yes," I shouted. "Sausages from the belly. Good idea." For the second time in my life I was grateful to be free from a confrontation over a joint of meat.

But I've got the hang of it now. The meat comes back from the butcher, frozen and labelled, half a pig to a box. I've often thought that if I had a spare afternoon I might stick one back together again to see what it looks like. But now I can use my little knowledge to take my revenge. I like to question customers about how they "prefer their loins".

"On the bone is delicious," I murmur, "but rolled can be quite an experience." I tried this line on one of my regular women customers but my wife overheard and has put a stop to it.

I've had only one complaint so far about the countless joints, chops and sausages that I've sold. A woman thought there was too much fat on her chops. Now, some butchers might take that lying down, but I have the advantage of not only being there when the chops go over the counter but when they come into the world, and if someone thinks they can moan about a bit of fat when our beloved sow has dedicated months of her life to rearing it they have another think coming.

I took the Basil Fawlty line: "Why don't you just cut it off?" I asked, sharply. It had clearly never crossed her mind: I suppose she resented paying for the fat she was going to throw away, but I wonder if she would ever ask a supermarket for the cost of the cardboard box her biscuits were wrapped in. I suspect she is one of those people who think fat is bad for you. She doesn't understand that meat without fat has no flavour.

Fat may or may not be bad for you: you can leave it on the edge of the plate, or live dangerously for once and have a nibble. But I'm not going to argue with her. She's a troublemaker. She can get her loins rolled elsewhere.

* * *

So there I was, forking muck on a balmy spring afternoon, arm muscles settling into the gentle rhythm of the swing of the fork, mind unwinding under the hypnotic influence of repetitive work. Organic farmers like muck, and are always happiest close to it. I was very close: I could savour every nutritious forkful, inhale each pocket of invigorating gas ruptured by my fork and, when pausing for breath, see newborn lambs at play, cows ruminating, fields becoming ever greener under the warming sun.

Then the precious moment was shattered. The woman came nosing into the farmyard, having spied the lambs. I had spotted her earlier, striding down the lane with the air of someone who owned the place. The only people round here who strut as if they own it invariably do not — except at weekends. "How pretty. Aren't they lovely?" she cooed. Like all proud fathers, I fell for the flattery. "If you want to see more lambs," I offered, "have a stroll up to the meadow."

I expected a word of thanks, but instead got a mouthful. "What *are* those?" she asked, pointing in disgust at the growing piglets.

"Pigs," I replied.

"But what sort?" she snorted.

"Large Black pigs."

"Well," she said, "they don't look very large to me."

Muck-flinging dulls the reactions, so I was unable to wither her with a barbed reply.

"Why do we see all these pigs in fields these days?" she continued. "Have you farmers just discovered bacon?" I opened my mouth, but no words came. She strode off, heading for the lambs, no doubt to frighten them as well. I think I remembered to warn her about the electric fence. Ah well, perhaps I forgot.

Still bearing the scars, I was wary when the next visitors turned up. It was a party of schoolchildren, and it is well known that no creature can cut you to the quick with the precision of a child. But I was pleased to have them, and their headmistress was relieved, too: taking children on farm visits these days must be a near impossible task. At some stage they have to learn that eggs come from hens and sausages from pigs, but you could hardly expose six-year-olds to battery chicken units or intensive pig-fattening sheds. Better they tickle Alice's ears or hunt for stray eggs. Modern farms are not safe places for children. The machinery is too unforgiving. Like an old music hall turn, I have a set patter for school visits. I start by showing children the harness, choosing the biggest horse collar and asking if any of them would like to try wearing it round their necks, as the horse has to. None of them can even lift it. Then I tell the smallest child to walk through it, which they usually can, without stooping.

Then I say I'm off to get the horse. By now they are bursting with anticipation, expecting a cross between Black Beauty and Nellie the Elephant. I pick the biggest horse. The children gasp; a carthorse close up is an awesome sight.

I remember a letter from a woman whose little boy had seen his first Suffolk Punch. The lad stared at the big brown horse, sensing its might and majesty, overcome by the shimmering brilliance of it, and said: "He looks like the sun."

You may think this is all sentimental twaddle, an

educational diversion, but if farmers want to be loved again they would do well to follow my example and start working up an act. As my nosy woman visitor demonstrated, no amount of public relations is going to convince hard-bitten adults that farmers have any good in them. Our only hope is the children. Anyway, children are always worth encouraging just for the thank-you letters and poems that arrive a few days later.

> It was lovely to see,
> Close to me,
> A Suffolk Punch
> Which had just had its lunch.
>
> He lived on a farm with some Red Polls
> And a family of pigs as black as moles.
> There were sheep too in a pen.
> I'd love to go back — but when?

A lot sooner than some I could mention, is the answer.

I know an old farmworker who lives alone in a cottage by a marsh. He lives simply, cooking his meal in a smoky old pot on an open coal fire, and when he is not boiling up food for himself, the pot goes on bubbling with potato peelings for the hens. His garden is neat and ruthlessly productive: beans, cabbages, spuds and doughty sprouts thrive there. He does not waste space on flowers except for a row of sweet peas, which he hands out to his lady admirers, who are numerous. He has been a friend of mine for a year, but this week he was the answer to my prayers.

My problems began a fortnight ago. At about four o'clock

one morning, I sat bolt upright in bed and announced that I was going to have to cancel the seed potatoes. Not only that, the cows were going to have to go. It was all becoming too much. Then, sweating with anxiety, I remembered there were 10 tons of manure to spread, meadows to be harrowed, pigs to be moved. "I'm cancelling the potatoes," I shouted again, this time with a sob in my voice as I thrust my head into the pillow in the hope of sleep. But there was no relief. Executives would call this a nervous breakdown and bring in therapists, but I suspect that such sleepless nights are a regular feature of farming life. Anyway, I couldn't cancel the potatoes because they were on the way.

Inspired by a holiday postcard, I decided perhaps I ought to say my prayers. The postcard, a folksy affair from the Austrian borders, showed "Saint Isidor, the Farm Labourer". He was a talented ploughman and devout man of God, so highly regarded that while he was praying every day, angels would come and do his work for him. The painting shows him kneeling before his church while a beaming angel in a nightdress ploughs the land behind.

My wife, who has custody of the *Dictionary of Saints*, tells me that, being a typical farmer, St Isidor stoutly denied he had any help. But I have learnt humility. I admitted I needed an angel, said my prayer, and soon fell asleep.

I am wary of admitting to visions, lest I be bracketed with the reborn turquoise tracksuit brigade, but I awoke with the name of the old man of the marsh on my lips. Thank you, St Isidor. I can't imagine why I hadn't thought of him before. He would make an ideal potato-planting companion. Like much manual work on farms, potato planting has a biblical simplicity about it. But it is slow, tedious and back-breaking, and if there is any way of getting a little unbiblical jollity into it, the opportunity must be seized.

I ploughed the furrows while the old boy filled the baskets and hauled the seed. He started at one end, I at the other, and when we met in the middle there was always a cheery tale to be told. There was the story of how he chopped off the end of one of his fingers, took it to the surgery to have it sewn back on but was told the doctor would be out for at least an hour. "Well, I weren't waitin' that long," he said, "so I chucked it away and went home fur m' tea."

Tales told, we would stoop once more to the potato planting. From a distance, a passer-by might think we were bent in prayer. But only St Isidor would have recognised which of us was giving thanks, and which was the labouring angel.

The earth moved for me this week. At the beginning of May we release the carthorses from their winter captivity and give them the freedom of the meadows. They have spent six dreary dank months living in their stable by day (when not at work on the fields) and by night they have rested in a strawed yard, sheltered from the cold in a lean-to, and munched dry hay. They do not complain. In fact, if I were to turn them on to the barren winter meadows they would stand at the gate and plead to be brought back in.

By April they feel the spring coming. Instead of plodding aimlessly round the yard at night, they stand sniffing the air, sensing the rising of the sap and the succulence returning to the grass. They look longingly over the gates and wait for their freedom. Some years it comes earlier than others, but this year, with an acute shortage of winter rain, persistent cold winds and no artificial fertilisers to speed things up, our meadows are slow to grow. Liberation has been a bit late, but when it comes it is a moment to savour.

At first they are nervous, but as soon as they see the grass and the first of their huge feet hits the meadow they fling their heads down with the force of hammer on anvil to bury their noses in it. Their great teeth rip the grass from the earth, and I swear their eyelids flicker in ecstasy.

This state lasts for only a few minutes. Such is the delight of the fresh grass that they must celebrate in movement. One horse will start to trot round the field, flinging its head high. This will distract the other and together they will trot, faster with each circuit. One will turn on the spot and kick out playfully at its mate, who will brake sharply with a slither and gallop in the other direction. Then they pause, breathless, until overcome by the headiness of the occasion they charge again, probably in the other direction. The scene can appear violent, but if you know the horses you can tell their bodies are coursing with delight. It can last for half an hour: biting, kicking, bucking, galloping, munching, playing.

Then comes the collapse. To watch a horse the size of a Suffolk Punch fall to the ground is to see a heavyweight boxer turn suddenly into a ballet dancer. It happens very slowly. Their heads droop as though they are going to faint; then their legs go at the knees as if turned to jelly; but they are in controlled descent, for as soon as they reach a few inches off the ground they shift their weight to one side so as to land their bellies with hardly a bump. A ton of horse has come to rest. Then, with a mighty heave and grunt, they twist themselves on to their backs and rub and rub and rub until some climactic satisfaction overtakes them. Their insides must slosh from side to side, for the accumulated gas produced by the rapid intake of fresh grass comes bellowing out of their rear ends with a force sufficient to wake the dead in the next parish. While one horse is indulging in this massage, the other may still be at the gallop.

This is when the earth moves. You can feel it several fields away. You hear a rumble of galloping hooves, punctuated by the rasping report of a gaseous escape. You can keep your dawn chorus and your cuckoos; these are my harbingers of spring.

This is not the time for introspection, however. The grass is growing, showers of rain are forecast. And I do believe I can feel a warm wind on my cheek. But perhaps it is only the other horse who has taken to his back and is saluting the turn of the season with a trumpeted fanfare that only a rolling carthorse can emit.

CHAPTER SEVEN

Hay Fever

Summer 1991

Hanging in a dark Suffolk outhouse, over a smouldering fire, is my very first ham. Every day that it absorbs the pungent, preserving flavour of the oak smoke takes me one day nearer to a farming dream fulfilled.

Farmers and countrymen of the old school, of which I am a disciple, knew the value of bacon and ham. Cobbett wrote, "A couple of flitches are worth fifty thousand Methodist sermons. They are great softeners of temper and promoters of domestic harmony." At the moment, a little good temper and domestic harmony would fit in quite well to our farming scene. I am furious that the much-needed rain has turned out to be carried by Arctic winds which are preventing any kind of spring growth: it is making me very bad-tempered. The domestic disharmony stems from the fact that my smouldering glumness is casting a pall over the household. We are living, as it were, in an emotional smokehouse. A good ham tea is clearly what we all need. If that fails, we'll try the Methodists.

The idea of curing a ham has been in my mind ever since the first litter of pigs was born; and so when they were eventually sent to be butchered, I asked for one leg to be left whole and on the bone. I then set about the traumatic process of learning to cure it. Curing is a preserving process, achieved by liberal use of salt and made subtle by the addition of spices, herbs, or more powerful flavourings like beer and treacle. But how?

My collection of aged books has much to say on curing, but they cannot dissociate it from the less savoury aspects of extracting food from the carcass of a pig. I remember a fine description of the technique required to scrub clean the entire outside of the pig's intestine and then, by inserting a stick (specially designed for the job) into one end of it and executing a kind of Wimbledon back-hand-flip, to have the colon snaking around the room and turning itself inside out. You could then scrub the insides clean. Not the sort of thing for a modern little farmhouse with Laura Ashley curtains. I wanted something simpler.

My answer came by accident from a remaindered book dumped in a bin on a railway station bookstall. It confirmed my suspicion that the old methods of curing required much greater applications of salt than would be acceptable these days. It made for long-lasting bacon but may well have shortened the lives of those who ate it. I settled for a "Suffolk sweet cure". I took a black plastic dustbin and sterilised it; I gave it sugar, black treacle, sea-salt, saltpetre and topped it up with six pints of foaming Suffolk ale. Like a sailor going to the watery grave of his dreams, the pig's leg slid gently into it. And there it has stayed for six intoxicating weeks. I have stirred it, and sniffed it, and hauled it out for an occasional inspection, with the family standing round respectfully. When the book said its time was up, I took it round to our local smokehouse.

I might mention that this is no ordinary smokehouse. It lies behind a small shop in a village not far from here and might be thought an undistinguished establishment, were it not for a Royal crest above the door. From these humble premises, hams for the Queen Mother's table are smoked and cured. We are rubbing, if not shoulders, then at least hams, with the nobility.

When the ham comes home and the first slice is tasted I shall report fully, but I can say with some confidence that it is likely to be superb; for although our ham is as yet untried, we have been eating and selling our own pork for some time. Palates jaded for years by bland chain-store pork have suddenly revitalised at the taste of ours. One family gave up pork ten years ago because they could not taste anything in it. They are now hooked, on ours. Cooks who have tried every dodge to get watery mass-produced pork to crackle when cooked, find that ours bubbles and browns to a juicy crunchiness. I could go on, but as I am not in a position to offer you all a taste it would be unfair.

Modern animal farming boasts of its yields, its economy, its success in breeding animals to streamlined commercial shapes. It sells its meat not on flavour but on the virtues of leanness and tenderness and tidy packaging, as if the standard of perfection in food was the Fish Finger. No wonder there is so little soft-tempered domestic harmony around these days, and so many evangelical sermons instead. William Cobbett and I, and the Queen Mother, we know better.

I challenge you to view my field of newly sown mangel-wurzels, and all the May-time buddings disfiguring it, and still find it in your heart to greet any of them as "darling". We have a fine crop of unwanted thistles and mayweed, docks and nettles, but of desirable mangel-wurzels we have none. The mangel seeds, unnerved by the cold and dry start to the year, have been shivering under the clods of earth since they were sown in mid-April, not daring to sprout. The weeds, on the other hand, have had an entire winter's dormancy in which to refresh themselves, and the mildest of excuses has them leaping into life.

In conventional farming it wouldn't matter, as a cunningly targeted chemical would eradicate the intruders and leave the crop to flourish. But we are attempting organic farming and are supposed to be more sympathetic to the patterns of Nature. Alas, she is not always sympathetic towards us. Weeds are her blind spot. You might think that all the effort we put into building fertility with natural manures and fertilisers might charm her into letting us off weeds for a season or two, but if you could see my mangel field you would know what a bitch this woman can be. Creeping thistles, clearly the invention of a vindictive mind, are my worst enemy. They have been biologically programmed so that their instinctive reaction when faced with death is to reproduce. If you take a hoe and slice off their heads they simply spit in your face and send up several shoots more vigorous than the last. There is an old farming proverb which goes:

> Cut 'em in May, they come next day.
> Cut 'em in June and they'll come again soon.
> Cut 'em in July and they're sure to die.

Which is all very well if you can wait until then, but I fear that by July the mangel-wurzels will have given up the unequal struggle.

After a particularly depressing meander through this field I retired to my aged farming tomes in the hope of finding solace. But the *Farmer's Cyclopaedia* of 1823 warns: "They . . . maintain so pertinacious a hold upon spots where they have taken root to be very difficult of extermination; they, in all instances, so facilely and multitudinously scatter their winged downy seeds that one free growth on a slovenly managed farm would propagate them over many square miles." Cue the sleepless nights for the slovenly farmer.

At least the thistles are not a problem of my own making, which is more than can be said for the kale stumps. Last year we grew a bumper crop of kale. By Christmas we were cutting it and feeding it to the cows, who clearly were unable to keep up with the speed at which it was growing. By March it was getting desperate so I offered the entire field of it to a shepherd with a couple of hundred hungry sheep. From fields away could be heard the sound of leaves being torn by the chisel-toothed flock. What a few cows had failed to do in a season, the sheep managed in a week. Except that they didn't eat the stalks. These remain upright in the earth like a forest of lolly sticks. I had to get rid of them. I tried the pigs, thinking that their ever-rooting snouts might relish a dig down to the kale roots, but they merely chased between them in a piggy slalom and left them unscathed.

I called in my neighbour, Farmer White. He turned up with a battery of mechanical choppers and diggers, and roared up and down until nightfall. It was only a partial victory: dawn revealed defiant stumps standing as proud as soldiers, ready to fight another day. I advanced with two horses and ploughed them under, which is what I should have done in the first place. Then I looked across at the thistles, wishing for as swift a solution. I retired again to my farming tomes. The first I opened bore the following inscription, which I believe to be biblical:

> I went by the field of the slothful, and by the vineyard of the man void of understanding. And, lo, it was all grown over with thorns, and nettles had covered the face thereof . . . I looked upon it, and received instruction.

If anyone knows what the instructions were, I would be pleased to hear from them. But make it soon. The thistle grows ever higher and down below, the kale stumps may be getting ideas.

This is not a plea for sympathy but I am in some pain. I hope it is a passing ailment but I fear that what I am suffering is all part of my transition into being a farmer. I have been a landsman in my mind for several years now, the body is only just catching up.

My feet are the problem. Since the middle of March the work on the farm has called for more trudging than I have ever done in my life. Now that the stock have been turned out to pastures I have to walk miles every day to check on them. I also walk up and down, over clods, behind ploughs, harrows, seed drills and cultivators until the leather soles on my boots are screaming for mercy. Until now I have felt invigorated by it, but recently I have started to walk in a bow-legged hobble. You might think I had nettles in the bottom of my boots; in fact my tender soles are feeling every stone and pebble.

In the meantime, work cannot cease. I have three acres of land to roll and harrow today and that means at least a three-mile stroll. Yesterday I took horse and seed drill to sow kale; that was another two mile meander. Over the winter I've ploughed eighteen acres, which is but a small patch on the modern farming scale, but each acre requires man and horses to walk eleven miles.

I am not complaining but my feet are. Last week I took them to my doctor, who, alas, was on holiday. His locum flippantly announced that I was "probably sickening for something". Had he been a vet, such a casual approach would have got him the sack.

So I have been elsewhere in search of a cure and, in what has become the tradition on this farm when faced with a problem, I looked backwards for the answer.

By chance, a precious parcel arrived in the post a few weeks ago. It was from a kind reader of this newspaper who had been given a set of aged exercise books in which an old Sussex farm horseman had written his veterinary secrets. Scripted lovingly in a copperplate hand on lined, yellowing paper were potions and lotions designed to calm boisterous horses, liven up drowsy ones, and cure horses with "stale blood". I flicked to lameness. The author warned: "With a perseverance in perpetual drudgery, [the horses] are brought to a standstill where they are too frequently observed dying wretched martyrs to the horrid combination of hard work, whipcord, and poverty." It summed up my position entirely.

Tantalisingly, some scraps had recipes but no indication of what they were intended to cure, so I can only wonder what the effect of 2oz brimstone, 2oz tobacco and 2oz gunpowder might have been on some unsuspecting nag. Anyway, ingredients were clearly going to be a problem. One shopping list bore a pharmacist's stamp dated 1912. The card listed liver antimony, grains paridise, sweet fennel seed, sublimed sulphur, oil of swallow, oil of spike and oil of rhodium. None of them sounds likely to be sitting next to the Night Nurse on the chemist's shelf.

I decided in the end there were only two options. One was to try a widely recommended "physic ball". If they work on the horse, they might work on me. They are packed, according to the notes, with powder of antimony, Castile soap, ginger and aniseed, and bound with treacle. They are rolled into 2-inch balls, and blown down the horse's throat. Or mine.

The other option was to walk less. We do have implements on which you can ride, in particular our horse-drawn mower. The only snag is that the seat is fixed to the machine by means of a sharp-headed bolt placed to provide the maximum discomfort. Having walked half a mile to the field with the mower and horses, I wasn't going to walk all the way back for the cushion I had forgotten. I have paid the price.

One way or another there seems to be no escaping the pain of farming life, or the application of oil of spike. It is simply a question of deciding which end to apply it first.

When beset by personal problems, some turn to religion or literature and a few of the more desperate will consult an agony aunt for solace. I always go and sit in an old barn; and the more cathedral-like the building the more comforting I find it. But not for the reasons you might expect.

An ancient barn built from gnarled and naturally forked

timbers is one of the most inspiring structures ever conceived for such a utilitarian purpose as sheltering cows, corn, or machines. But my eyes are not drawn to the vaulted rafters or the thatched roof, rather to a cunning little gap often left between the inner and outer walls. This is where the treasures lie and the magic of the barn resides. To draw an ecclesiastical parallel, rather than gaze on the glories of the stained glass, if you want to see the light I suggest you look under the pews.

By strolling round decrepit barns I have, with the farmer's permission, collected indispensable items of gear for our farming enterprise. I have unearthed items of harness like wontys, dutfins, top latches and hame straps. Farmers think they are junk (which they are to anyone who doesn't farm with horses) but are vital in holding together our link with the horse-drawn farming era.

But this kind of treasure-hunting can have its heartbreaks. An old lady was showing me her father's derelict farm. I asked where the door in the end of the barn led. She couldn't remember. I turned the handle and fought back the nettles to find, hanging on the wall, a dozen sets of carthorse harness, carefully arranged at the end of the last day's work the horses had ever done. I reached to pick up a collar and it crumbled to dust. The bridles turned to powder if moved. I was too late.

But last week I was just in time to save a rare machine. In return I expect the rare machine may save my crop of corn. Beset by worries about the weeds, I responded swiftly to a call which suggested an old barn near here might be housing a "full stetch, steerage horse hoe". Let me explain.

A stetch, in Suffolk parlance, is the width of land sown

by the seed drill. This machine, pulled by horses, was the same size as the drill, and hanging from a movable bar were a series of hoes designed to be steered between the rows of growing corn. I could not have been more elated if I had found a Constable in the attic.

When I arrived at the farm the vibes were good. It was once very prosperous, with a huge farmyard and stabling for at least a dozen working horses. It took little imagination to visualise it littered with haystacks, or corn waiting to be threshed, and animals. Now it was silent: the big machinery had long since taken over.

So intoxicating was the atmosphere that I became convinced it was my mission to rescue this venerable machine, bring it back to life, drive it across the land once more, proving that there was much in the old ways of farming worth preserving. The hoe was the only method of weed control farmers could resort to before chemicals took over completely, so if ever there was a "green" machine this was it.

It was in fine shape, considering it had not been used for at least fifty years. The wooden wheels were sound, the axles turned freely, the hoes swung from side to side and still had a keen cutting edge. It was on my trailer and homeward bound before the farmer could say thank-you for the fifty quid.

The next day, I could not resist a return visit. This time, with the farmer's permission, I went alone. I strolled down the dusty nave of the great barn and worked my way along the transepts, looking. I heard the massive door open behind me and an old man appeared. I had met him before on my first visit, and knew he had worked here for fifty years starting in the Depression days of the thirties, ploughed for victory in the war and lived to see intensification overtake the fields, the hedgerows, and his way of life.

"Yer goin' hoein', then?" he asked.

"Is it easy?" I replied, somewhat nervously, having looked at the machine and considered the impossibility of steering a dozen hoes side by side through a field of corn where the rows are a mere five inches apart. Get it right and you rip out the weeds; get it wrong and you've sliced through this year's harvest.

"There ain't nuthin' much to hoein' when you get the hang of it."

"And how long does it take to get the hang of it?" I asked.

His eye twinkled. "I wouldn't say I'd got the hang of it yet." He chuckled and shuffled back into the darkness. I looked up at the vaulted roof, thought of the wide hoes and narrow rows of corn, and said a prayer.

I am having an identity crisis. I am no longer certain what kind of farmer I am, or indeed whether I am a farmer at all. Most farmers are men with hundreds of acres and battalions of machinery. They don't struggle up to their orchards with buckets of pig grub, as I do twice daily. They wouldn't think I was a farmer. If they bothered to give me any consideration at all they would think I was a crack-pot, or more likely an irritant who spreads this organic, nostalgic nonsense: a pest they wish they could spray against.

But some farmers would be more generous and call me a part-timer. This is not quite right either, for although it is true that I earn part of my living in other ways than off the land, it is that side of my work which feels part-time.

Actually it comes as rather a restful treat after the struggle with the pig buckets.

If they really wanted to needle me they would call me a "hobby farmer". I met an estate agent who owned 600 acres and reckoned he was a farmer and I was a hobbyist. But he farms his land from behind his desk, and while he is driving around in his Japanese jeep waving detachedly at his farm, this so-called hobby farmer is out mending the fence the sheep are destroying. He'll have scanned his balance sheet, checked his feed conversion ratios, and glanced at the barley futures on the city pages while the alleged hobbyist is still struggling to get the rust off his aged Lister Blackstone swath turner before hay-making.

I have now come to the conclusion that I am not any of the aforementioned; in fact I am one of a rare breed. I am a peasant. A trainee peasant, at least. I just want to wrap my modest parcel of land around me and get on with it. It makes for an insular life with peculiar problems. Machines, fertilisers, and medicines now come in parcels too large for us to handle. I have a cow with lice: the smallest bottle of treatment I could buy was sufficient to dose fifty animals and cost over fifty quid. I think the cow was as shocked as I was, for when I told her she stopped itching immediately and hasn't rubbed herself since.

Small-scale farmers don't seem to fit into anybody's scheme of things. I glanced at one or two training schemes that were on offer, but they were all about operating machines that I had never heard of which wouldn't fit through our gates. Or else they were useless. One course offered farmers a day's tuition in "Improve your Telephone Selling Technique". Peasants are ahead of that game already:

the other day a lady rang up to order a couple of joints of pork and in conversation I persuaded her to have a pound of sausages as well. I would have had her taking the liver, had I not spied a horse which had slipped his halter and was heading for the wide blue yonder. There was another course darkly named "Stress Management for Farmers". It is not my idea of a good day out to spend it in a room full of morose, whingeing, 600-acre-owning estate agents.

Yesterday's post brought an answer to my problems. The sensible people of Norfolk have formed a "Small Farmers Training Group" and offer lessons in mundane but vital tasks like "catching sheep" or "handling cows". Their lecture on farm machinery is, I notice, to be held in a scrapyard. I feel that the Norfolk Small Farmers and I are on the same wavelength.

I decided to book myself in for a blissful afternoon entitled "Sharpening Tools". If you have ever tried to cut a verge with a blunt scythe and found that your attempts to sharpen it only make it duller, you cannot appreciate what a lifeline this could be. It was the final line of the billing that clinched it for me. It said simply, "Save money and temper". It was my sort of stress management.

I rang the helpful secretary and we discussed my problems of nationality, coming from Suffolk and wanting to attend a Norfolk club. We decided I would pay a cash penalty and would then be welcome to attend whatever I wished. "Sharpening Tools, please," I blurted. There was a deathly silence through which even my blunt scythe could have cut. "I'm sorry, we've had to cancel it. Not enough interest."

So I stand alone, not part-time farmer, hobby farmer or even small farmer. If the entire county of Norfolk cannot summon up enough enthusiasts to share my desire to put a sharp edge on a piece of steel, I must be a rarer breed than I think. Perhaps I should put bars around the farm to protect myself, and call the place a zoo. The 600-acre boys could come and push bananas through the bars at me. They'd like that.

My field of barley is awash with crimson poppies and a joy to behold. I suspect that as a farmer I should view this as some kind of failure: I know a chap who let a field go fallow one year and the result was a display of poppies so vivid that people came from miles around to see it. He was very embarrassed. Having spent years employing chemicals and technology to control his weeds, he felt he deserved more than to have the hardy poppy thumb its nose at him.

But it is not the poppy's powers of survival that have endeared them to me recently. They are also the symbol of remembrance and gallantry, for which this has been a fortnight I shall never forget.

The drama started at midnight on the longest day of the year. Just as I was getting into bed, I heard a squeal from the pigsty, so faint that it might have been a slight movement of a rusty hinge. Except that it had a blend of bewilderment and frustration which I have learnt to recognise as the alarm call of the newly born piglet. Alice, the Large Black sow, already a mother of twenty-five, was at it again.

She rarely makes a mistake in giving birth, so I was slightly concerned at the whimpering. Generally speaking, you can bet that no sooner are piglets born than they set off purposefully for the nearest teat, which they find with hardly any trouble. But one flash of the torch into the dark sty revealed Alice's major miscalculation. No doubt in order to enjoy the cooling breeze around her rear end while at the same time giving her snout the inner warmth of the sty, she had plonked herself down in the doorway. It is quite a wide door, but she is a very wide pig and consequently when the newly born set off in search of something to suck, they found themselves impeded by a firmly wedged Alice. It was like expecting babies to cross the Black Mountains for their first taste of mother.

I employed a technique not used in obstetrics in this country for some time, and shouted, "Get up, you daft bitch!" So shocked was this *grande dame* at being addressed in such a manner that she heaved herself on to her feet, ambled inside, and settled down again without even bothering to give me a grunt. That pig has a withering way with her silences.

Next morning we had eleven piglets and a major problem. Alice had a teat the size of a cricket ball and just as hard. It felt hot and looked tender. The vet confirmed infection and warned that other teats might be suspect too. He gave her an antibiotic but doubted whether Alice would be able to feed her litter. Only nine teats among eleven piglets spells trouble. Kindly, although it was a weekend, the vet drove off to find us a sack of "Sow Milk Replacer". But when I saw the gloomy package and its list of contents, it read like a food additives horror story. It had antibiotics, growth promoters and a sinister ingredient described as being "denatured according to EC regulations". (Aren't we all?)

I am not against giving drugs to animals to cure or to save life, but I hardly felt these healthy hungry piglets deserved a pharmaceutical belt round the ear at this early stage in their lives. So I discussed it with my wife, and we came to the conclusion that Alice might conceivably hold the solution to her own problem. We went back to the sty and told the old sow that we were placing our full trust in her.

But we didn't abandon her. We paid hourly visits, sometimes bringing bunches of fresh clover from the meadow like relatives visiting the sick. Under the influence of her medication, poor Alice was clearly under the weather. She didn't move much except to eat and then with only half an appetite. As for the piglets, we expected one morning to find two of them dead and some kind of natural selection to have taken place. Alice would know best.

But a week later, all eleven are thriving. I do not know how she has done it; perhaps she has devised some intricate rota system, for none looks underfed or sickly. When my wife had a fit of panic and decided to bottlefeed the smallest, it resisted arrest and spat out the teat with such force that she retired shaken.

Alice, too, has regained her strength. She is a heroine, whose determination has seen her litter through. Whatever happens now, it has been an act of bravery and every year when the poppies are in flower, we shall remember.

I read that a team of Japanese scientists has set up camp in a secret location in southern England with a view to discovering the truth about crop circles, those mysterious flattenings of standing corn which leave regular, geometrical patterns. These are believed to be caused by either an inter-planetary form of communication, or a load of drunken young farmers having a long-running joke at our expense. The Japanese clearly take this seriously, for there are no less than nineteen of them hiding in the corn waiting for something to happen.

I have to tell them that they are looking in the wrong place. While I was strolling through my field of oats the other day, I noticed patterns of which I had not been aware before. Those of a vivid imagination may insert sinister music at this point; those who are commercially minded might follow the example of the West Country farmer who, within hours of finding circles, had the farm open to the public and the T-shirts printed.

But before you start getting together a coach party, I ought to explain that our crop circles are not quite as spectacular as some. They are perfectly round, as good crop circles should be, but instead of being a hundred yards in diameter they are a mere couple of feet. In each circle the earth is bare and parched, but around the perimeter the corn is lusher than in any other part of the field.

I could keep you in suspense but it would be unfair. Each of my crop circles marks the precise point at which the carthorse paused to relieve himself when we were planting the corn. I assume the potency of horse urine is such that it has a poisonous effect on young seed, but as it spreads out from the original spot on which it was deposited, it becomes diluted and fertilising. Either way, it seems hardly worth spending a night huddling in the oats to find out. It is either aliens or horse piddle, and I know which explanation I believe.

If these eminent Japanese scientists really wanted to see one of the mysteries of the natural world they should have been here last week when my old friend the marshman came to "chop art mangels". He is a retired farmworker aged seventy, an old soldier, and boasts he "only grows one marrow a year, but that's big 'nough t' keep me bike in". You have met him before in this column when he saved me from despair during the potato planting. This time his mission was to save the mangel-wurzels from suffocation. Mangel seed is so vigorous it sends forth three or four plants from each seed planted: if allowed to grow in such an overcrowded state it becomes a crop of useless, spindly, delinquent roots instead of a satisfied community of stout mangels as big as footballs.

I doubt my friend has ever bothered to master any mechanical aid, but give him a simple hoe and he can make it sing. Like a fussy snooker player with his cue, he carries his own hoe and will refuse any other. He tunes it with a swift rasp of a file to give it an edge, and to ensure a grip he spits into the palms of his hands. Then the ballet starts.

He stands back a row or two from the one he is attacking and takes what may appear to be a casual swipe. It is carefully calculated. That single action removes half the unwanted plants. Then, with a twist and a nudge, he closes in, swipes again and miraculously manages to remove individual weeds and give the soil an invigorating stir without altering his grip or stance. Then he shuffles a pace down the row and repeats the action . . . swing, swipe, slash, stir, shuffle.

He does this for six consecutive days, rain or shine, sitting in the ditch to eat his cheese sandwiches and refusing all invitations to the shelter of the farmhouse kitchen. And at the end of it all you have a field of vigorous individual plants revelling in the freshly stirred soil and growing before your eyes. That's my sort of magic. But he's never going to be famous for it. No oriental delegation is going to try to unravel the secrets of his art.

Anyway, he and I have developed our own theory about these corn circles: they are the product of a giant, inter-galactic horse, and if it should relieve itself while inscrutable eyes are bonded to their binoculars, we can only say it is as much as they deserve.

If you have written to me lately and received no reply, I can only apologise and hope you will understand how the rawness of a farming lifestyle dulls the social senses. For example, I owe a sincere apology to two elderly women who were sitting near me in a chintzy teashop recently and who were unfortunate enough to glance over my shoulder.

I was taking an afternoon off and had bought a manual on pig medicine. It is by far the best book on the subject I have seen, giving lurid details of such diseases as vibrionic dysentery and bowel oedema. So engrossed was I by the chapter dealing with mastitis (from which our sow, Alice, suffers) that I hardly noticed the two women remove hurriedly to a table at the far end of the room. Perhaps a vanilla slice and a Technicolor close-up of pigpox are not an appetising combination.

But to return to your letters. They are a joy, for they invariably encourage or advise, and even those that warn do so in a helpful way. I have had advice on my aching feet (fully recovered, thank you) ranging from thorough applications of coal-tar soap to complex blends of vitamin B. A letter from an old farmer told me of the days when he was a child so small he was able to walk beneath the bellies of the carthorses as he spread the straw. A story or a word of encouragement is as valuable on this farm as a forkful of muck.

Some letters do more than fertilise: they are seeds in themselves. This week, a letter from a man, aged seventy-one, in Darlington, blossomed into a grand dream, which I now lay before you.

Moved by my accounts of battles with weeds, Mr Dedman reminded me that the reason for growing roots (we grow mangel-wurzels and turnips) was partly to clean the land of weeds. The seeds were planted in neat rows, the horses dragging hoes between them to kill any sprouting weeds. By the end of the season you had vigorous turnips and dead weeds. As he advised me, "the more you hoe, the more they grow!" Timely encouragement in a week in which we have been bent double in the mangel field, smiting the soil. But apart from the cleansing effect of hoeing, he went on, there was a further incentive to do the job properly: you had to satisfy a wartime government official known as the "Thistle Man".

This spiky character, who usually arrived on a bike, was charged with inspecting farmland for weeds. This must have been very irritating. In order to please him, farmers would organise a kind of military march. As many "conscripts" as could be mustered would be lined up along the field edge, three feet apart, each armed with a hoe. When the order was given, they would march forward with orders to kill every thistle within reach. Captain Mainwaring would have been in his element. If you think this was excessive, it is as nothing compared with the time, effort and fuel being expended by our council (and probably yours) in a military-style operation of mowing the verges of country lanes. Without the Thistle

Man to direct them they trundle up and down the lanes scything flowers, grasses and weeds in what I can only assume is the pursuit of tidiness. Where it makes a road safer, I am all in favour, but mowing for mowing's sake smacks of suburban rather than rural inspiration.

I became even more confused when I drove along a motorway and found miles of verge strewn with deadly ragwort. This is a vicious little weed that is poisonous to animals and lethal if it happens to be made into hay. There was every excuse for destroying that particular growth, but clearly nobody intends to. Instead, they thunder pointlessly up and down lanes making cowardly assaults on innocent cow-parsley.

We need the Thistle Man back to organise their misdirected energies and make sure the right weeds get zapped. It was the advent of chemically controlled farming that put the Thistle Man out of a job. Now that we are living in greener times his day may come again: an essentially British, bossy, useful figure, teaching us to leave harmless weeds alone and stamp out encroaching poisonous foreign ones. Who could do the job? Who is free, patriotic, and has the moral authority to make grown men stand in lines and do as they bid? Of course, there's no reason it should be a Thistle *Man*.

The government is urging farmers to diversify. This means it wants us to do anything we can think of with our land, except grow food. Any scheme must generate income for the farmer, cause no public nuisance and have an element of friendliness towards the environment. If, at the same time, it covers the government's embarrassment at having allowed agriculture to slide from a noble calling to little more than a rural nuisance, so much the better. May I now unveil my plans, which I trust the ministry will applaud?

I hear that the smart thing for executives to own these days is a personal trainer. These female trainers dazzle from hair to toe; they are as bursting with fitness and energy as our sow Alice is stuffed with pig-swill. They leap and bound into the humdrum lives of executives, leaving behind them a trail of sweat, sprain and stretched stomachs. I may not have the glamour (or the leotard) that normally goes with the job, but as I was wandering the farm the other day I realised that we have more to offer than any of them. So I think I shall turn the farm into a health resort for harassed executives who need to get away from their desks. Or indeed, from their personal trainers. Of course, we'll need to smarten things up a bit. I thought we might build a reception desk in fake teak, just like the real health clubs. I might knock up a little shed for it out of some old railway sleepers. The spot I have in mind is where the, er, liquids drain from the cattleyard, but if I put the receptionist in wellies it shouldn't be too uncomfortable. Anyway, I shall need a hardy lass. I am rather hoping to recruit a retired air hostess, because I feel only she would be able to cope with the moment when, after signing in, our clients are issued with lengths of string and advised to tie them around the bottoms of trousers at all times. (Our rats are unused to the odour of Bond Street talc, and may prove curious.)

This lass will also have to break the news to clients about their personal programme. "Good morning, Sir Ralph. You're on sheep-dagging today, aren't you?" Or, "Sir James, you'll find twenty piglets in the sty. Yes, some of them are quite big. You'll be doing ear-notching this morning."

Naturally, we must develop a graded fitness schedule. Upper abdominal development is fostered by an exercise in which you stack bales of straw until you collapse exhausted, only to have an escaped cow knock them all down again. We play this game often. Then there is the primal therapy of thistle-murdering. I can just hear those squeaky-bright, disingenuous tones echoing round the farm — "One . . . two . . . three . . . HUP! And HUP! And swing that hoe!"

For wrist muscles there is an exercise involving the blocked farmyard drain. For this one we lie down . . . and reach . . . and grab . . . and pull . . . and throw . . . and reach and grab . . . "Pull harder, Mr Leigh-Pemberton! We want it all out!" Then there's the mangel-wurzel harvest, where we bend and grab, and slice and throw, lifting the mucky roots from the slimy soil, skimming the tops with a knife to fling them into the cart. We aim to lift five tons an hour. Arnold Schwarzenegger would hardly get through the first ten minutes.

Of course, it would be wrong merely to exercise the body and not build character. So I am developing a mental stimulation game. This involves entering a field of sheep, whereon I announce that one in particular looks as though it has a headache, and ask our guests to catch it. Any swearing will result in the withdrawal of privileges, such as food. If they get too good at it I shall give them Flash, the sheepdog, to help, having first ensured that he is in a filthy temper.

Then there's the game called "Load the pig into the trailer." A good one, this, for over-bossy executives who think they can get their own way by throwing their weight around. If anything gets flung around when the pig decides the trailer is unappealing it is the student. "Oh, sir, are we upside down in that trough again?"

If all this sounds demanding, it is. Executives say of their personal training sessions that it is the first thirty seconds that are the worst. Of the programme I have in mind, fellow farmers tell me it's the first thirty years.

I have cocked it up well and truly, and am rather proud of having done so. I refer to our hay. In a normal season, hay is made by mowing grass and allowing the sun to dry it. A good breeze helps, too. When it is dry on one side, you flip it over and allow it to cook gently on the other. After about five days it becomes hay and you put it in the shed.

When the going is good it is a pleasant enough business; especially if you are farming with horses, for the flipping-over of the swathes of drying grass brings out heady scents that would never penetrate a tractor cab.

But June showed no sympathy, and it was becoming desperate. I was hoping to make hay out of my red clover crop which, although highly nutritious when dry, is a soggy sort of crop when freshly mown and badly needs heat.

"What you wanna do, boy", said my consultant, aged eighty or so, "is do what we used t' do. Yer wan' 'er cock that up!" So I did.

The weather forecast gave a promise of three dry days. If we could get it cut and partially dried, we would be able to build our cocks and allow the rest to take place within them. Cocks are miniature haystacks, loosely built. If you made a big stack of damp clover, it would go mouldy; it is built around tripods with passages to allow the free flow of air, it can stand safely in the open for weeks. So the old boy told me.

On Thursday I got the horse-drawn clipper out of the barn and gave the Suffolk Punches a hearty breakfast. ". . . Expected to stay dry in the east till Saturday afternoon," said the forecast. "G'up," I called to the horses and the rattle of the mower cut across the valley for all of half a minute. The crop was over-thick after the wet June, and the knife had clogged. It took five minutes to free. "G'up," and we clacked along. Then we jammed again. I felt the three dry days ebbing away and hardly 50 yards mown. I adjusted the mower to cut higher. "G'up." One horse shot forward like a bullet and the other didn't move. The harness had broken. Back to the stable, hot and very bothered for more leather.

We cut three sides of the field and turned the horses to take the fourth. "G'up." They moved forward, but with hesitation and I don't blame them. Ahead lay a thick forest of Scotch thistles, each spine ready to torment a sweating horse's body. As we reached the clump the horses slowed, so the cutter jammed. It was lunchtime.

". . . With showers breaking out in the east by Saturday morning," the radio said. They'd moved the rain forward half a day! Lunch was grabbed and back to the field. We abandoned the thistly patch and set to on the rest. It was tea time.

". . . With rain, heavy and thundery overnight on Friday in the east." Again! With every sweating step we took, the weather forecasters were inching the deluge closer. Quite frankly, I didn't believe them. I rang a special number which gives a recorded weather forecast for farmers. It begins 0898 and my finger, still wobbly after a day on the mower, must have misdialled. I reached a panting lady who said, "Hello there, Big Boy. Does your wife know you're calling me?" I assumed this was another of Mr Gummer's farm diversification ideas, and rang off.

I awoke the next morning to find the sun blazing and the cut clover much drier. The forecast was for no rain till late that night. I took a horse and our vintage swath turner and turned every row so the damper underside was exposed to the sun. I sniffed the fragrance of the cooking clover, and scented victory in the air.

About four o'clock we set up our wooden tripods. The horse-rake dragged the clover into heaps and with forks we placed the heaps around the tripods, building them with care till they rose 8 feet from the ground, rounded on top but steep-sided to repel the rain. It looked like a primitive village: a sight not seen hereabouts for many decades, an ancient pattern back in the summer landscape.

We finished at half past ten, exhausted and aching, and I watched the late weather forecast. Rain was heading relentlessly our way — "Downpours in the east tonight!" Happily we went to bed.

It is now four days since we finished, and it has not rained. I guess somebody else cocked it up, too.

We have a serious outbreak of a virulent disease. It leaves you dizzy and bewildered. I am suffering from being over-advised. I reel under a hail of well-meant but conflicting nuggets of agricultural wisdom.

I had my first hint of this problem a couple of years ago at a farm auction, buying my first horse-drawn plough. I asked a couple of knowledgeable old boys if they thought the plough better suited to light land or heavy. They replied instantly and in chorus. One said "heavy", the other "light". Earlier this year, I was pondering whether it would be best to undersow clover seed with the barley or the oats. My adviser on that occasion said he wasn't certain but he knew where he might find the answer. The next day I had two telephone messages. One chap had rung to say "barley", another had called to say "oats".

Now that the haymaking season has at last arrived, I find the advisers are out again, hovering like a swarm of bees looking for somewhere to settle and cause irritation.

The first adviser took one look at the field and told me I was wasting time making hay in the first place. "That old meadow hay, that's not worth cutting. Not worth cutting, that isn't." He had a point. It was poor grassland, matted, knotted and weedy. But should I explain to hungry sheep in the depths of winter that there is no hay because the field didn't look quite tidy enough to cut? And what of the chap who, the day before, had told me: "Well, in my day, we never had better hay than came off them old meadows. Never had better hay!"

There was a period when I was contemplating ploughing that meadow and reseeding it with fresh grass. "You want to get a good vigorous mixture. Lots of rye-grass." I was on the point of ordering, but thought I might glance through the aged farming tomes before doing so. One book warned: "Rye-grass is not highly nutritious, will not fatten growing cattle and gives out in July and August. Other grasses are far more worth growing." However, of another grass variety called cocksfoot, he wrote ". . . our most valuable pasture plant, fulfils the highest wish of the humus builder, greatest yielder of forage, always forms the larger proportion of our most famous pastures". I decided on cocksfoot. I mentioned to a farming friend that I was thinking of sowing a cocksfoot meadow. He shook his head and spat.

So the rough old meadow stayed and grew and is now ready to cut for hay. While I was sharpening the blade of the clipping machine, an old farmhand dropped by. "Are you thinkin' o' clippin' that there owd me'dow?" I told him I was. "Well, what you want to do, boy, is set that clipper high so that's not getting tangled up in them roots. Set it high!" I would have done so if I had not remembered the advice of a rustic sage: "One inch at the bottom's worth three at the top."

The same chap asked me if I was going to stack it or bale it. I told him I was going to build a haystack. "In that case," he said, "you'll be all right 'cos that hay won't need to be quite as dry as if you were goin' t' bale it." I sighed with relief. The next visitor warned: "I've lost more ton of hay by stacking that too soon than I 'av b' leavin' it. Leave it till it's right dry." This was an important point. The moment at which the hay is carted is as crucial to a farmer as the grape-picking time might be to a vintner. Gather the hay when it is of just the right moisture and you guarantee sweet fodder. If it is too wet it cooks in the haystack and rots. I went back to the books. "Overheated hay is as bad as none at all. It will kill cattle, and ruin horses," warned one author. "Hay, baked in the sun till bone dry, is useless," insisted the next.

As I began to cart the hay another adviser appeared. "You ain't cartin' that today are yer? Weather ain't right. It be clung." "Clung" is a word I have heard before in Suffolk. It can describe a damp muggy day, or a lettuce too long in its supermarket film, or a spineless individual. I can't define it but I know what it means.

In the end, advisers advise but farmers must decide. I am carting the hay, pitching forkful after forkful on to the wagon. It is tiring work when you are swamped with well-meant advice. I feel like a drowning man, clung to a straw.

Farming, especially at the pace dictated by carthorses, gives a man plenty of time to brood. The slow, lonely plod up and down between the swathes of hay allows him to ponder on his grievances, and turn smouldering indignation into blazing umbrage.

For example, the morning I went to turn the hay for what I hoped would be the very last time before carting, I noticed in the newspaper that on the lunch menu at the G7 economic summit was a course described as "Suffolk Pork".

Well! What I want to know is why it didn't come from here. I brooded over this for the entire length of the field. I decided in the end that it was probably for the best, since it would have been so delicious that they would all have come back for seconds, lunch would have overrun and we would have been faced with the unedifying prospect of poor Mr Gorbachev pacing up and down outside while George Bush and Helmut Kohl argued over who should have the last bit of crackling. It was probably in the best interests of a new world order that the Suffolk pork did not come from our farm.

Having got that out of my system, another little niggle wormed its way to the surface; this one was also political.

At a recent meeting of European agricultural ministers who were arguing over the future of the community's farming policy, our very own minister declared that there was "no point in looking to the past for our answers", or words to that effect. If I cannot put pork on their plates, allow me to dump a parcel of papers on their desks.

This sheaf of moth-eaten documents arrived this week from a lady who unravels family trees. Genealogy, it is called. When she tires of family trees, she unearths farm histories, and has done a little digging into our parcel of land. Among the documents lay an agreement written in copperplate, which may prove to be the solution our political masters are seeking.

Before getting down to the agricultural nitty-gritty, it rambles through many quaint paragraphs of legal "appurtenances thereto belonging". Landlords did not simply let the land, it appears, and take the money; they required a detailed say in the way it should be farmed. In those days there was no agrochemical way of forcing crops out of impoverished soil: the landlords had an interest in keeping it sweet. So the farmer is required to "farm the said lands in an husbandlike manner and as follows: one fourth in wheat, one fourth in Barley or Oats, one fourth in Summerland [i.e. fallow], one eighth in peas and beans, and the remaining one eighth in clover."

Old-fashioned? Maybe, but at least it was a healthy way of farming from the soil's point of view. It built fertility and cleaned the land of weeds. It did not stipulate that "every other week thou shalt go forth with another 300 gallons of chemical and blast hell out of every living thing".

Crops were less heavy than today, but then it is surpluses which are at the core of our problems.

The old deed goes on: the farmer shall "covenant not to

break up any of the pasture lands without leave in writing". As far as the growing of corn is concerned he shall "sow clover, or other grass seeds, with the barley or oats".

Farmers of the old school will recognise the undersowing of corn with clover as a classical method of building fertility. Clover enriches the soil with nitrogen. Other little clauses insist the farmer must "consume all clover, roots and corn and return the manure to the said land".

This could almost be a document outlining modern organic farming, except that it was written in 1828. It insists on natural fertilisation and standards of husbandry guaranteed to leave the land in good heart. It was labour-intensive too, and we desperately need more jobs in rural areas.

So, as the world summits no longer have anyone at the table who seems prepared to press for a return to Victorian values, may I offer this solution to our leaders? If they want further details, I suggest they apply to our local Record Office, where the answer to their problems has lain for nearly 200 years.

CHAPTER EIGHT

Harvest and Hard Decisions

1991

There is a striking phrase in an old book I have been reading. Of the farmer it says that he ". . . is one that manures his ground well, but lets himself lie fallow and untill'd". This is true. Preoccupation with the relentless tasks of farming allows no time for improvement of the mind; or even space within the mind to let any thoughts flourish other than those concerned with the desperate business of getting seeds to grow and crops to flourish. I am now finding that entire cultural experiences, especially the more transient ones, have passed me by. Hence, while Madonna was parading her corsets, I was engrossed in my binder; and as for Pavarotti, I remember that when I heard he was to sing in the park, I only winced at the thought of all that luscious grass being trampled.

But I have not been living in isolation. I have formed my own allegiance: to men who unknowingly have given support during my first faltering steps through the farming year.

I am a fan, for instance, of Arthur Young, an agricultural

observer of the early nineteenth century. He took himself on tours of the shires and wrote a headmasterly, and not always complimentary, report on each. In his 1813 report on Suffolk, where we farm, he wrote of the harvest that "... they are more careful and attentive in many parts of the kingdom, in harvesting all sorts of corn, than they are in Suffolk". I blush at what he would have thought of our first harvest. Of haymaking he grumbles: "This branch of the farmer's business is but imperfectly practised in Suffolk." He may be an old misery, but when it comes to clues which might help recreate the old style of farming, he is a gold mine. I am going to sow grazing rye to give next year's lambs and ewes an early bite of green stuff. Young thought it was a good idea.

But if I turn to Young for fact, I turn to Thomas Tusser for entertainment. Tusser assembled his wisdom in a lengthy sixteenth-century verse, *Five Hundred Points of Good Husbandry*. For a scholar from Eton, Trinity, and the choir of St Paul's Cathedral, he has the poetic talent of the cracker-motto writer. For example, he describes the ideal farmer's wife:

> Good huswiues provides, eer an sickness doo come,
> Of sundrie good things in hir house to haue some.
> Good Aqua composite, Vinegar tart,
> Rose water and treakle, to comfort the heart.

But I revere Tusser. I treasure a horse-drawn hoe, a tumbril, and a seed drill which have covered that self-same land he farmed at Manningtree on the Essex/Suffolk border. This may be as silly as keeping a garter from the same factory that made Madonna's suspender belt. But as I trundle these old bits of gear around the farm, the verses of Tusser rhythmically bounce through my head. At haymaking time I was inspired to fork harder when I remembered:

> With tossing and raking and setting on cox.
> Grass latelie in swathes is hay for an ox.
> That done, go and cart it and haue it away.
> The battle is fought, ye haue gotten the day.

Try that one to music, Madonna.

As I strolled innocently into the kitchen the other day, I felt the same horror that must have gripped the sailor in *Treasure Island* when he opened the envelope to discover the black spot. Lying on the table, giving no hint of its latent menace, was a gently curving six-inch piece of wire. I hoped it might be the broken end of a wire coat-hanger; but when I saw a label hanging from it my fears were confirmed. I read the shakily written words out loud so that there should be no doubt. It said: "Binder Trip Spring from Gordon".

The wire was a vital component in our binder we shall use to harvest our oats and barley. I have been putting off the day when I would have to face this tangled mass of cogs, belts, chains, canvasses and flailing lengths of wood. I have been pretending that somehow this apparently incoherent machine would, if left alone in the barn, pull itself together sufficiently to be hauled round the field and picturesquely turn standing corn into sheaves. But the sight of the trip spring, a gift from my old friend Mr Sly, confirmed that the day of confrontation had arrived.

First I rang my benefactor to thank him for his gift, without which a binder is as much use as a jumbo jet without the ignition key. Mr Sly, you may remember, is the retired machinery dealer who unearthed this 40-year-old binder in the first place. But it turned out that the trip spring was even older than the binder. Mr Sly's brother, a Suffolk farmer who liked the security of a good set of spares around him, had recently died and in his desk drawer lay the spring, put away carefully for a rainy day. Only a man in the know would have spotted its true worth. Trying to buy one today would be like hunting for spares for Stephenson's Rocket. I was grateful to have it.

However, the most pressing problem was not the lack of a spring but a lack of coherence in the entire machine. I set a lad to remove the rust and solidified grease of five decades and give it a lick of paint. But who was going to lay his magical hands on this confounded device and bring it back to life? The phone rang. My prayers were answered. It was Farmer Jones from up the hill, inviting me to view his collection of farming antiques.

I spent a happy evening in his barn enjoying old forks and shovels, gas masks and bits of beloved machines long since departed. But when he showed me his father's old car, I knew that Farmer Jones was the man who could wave a recuperative wand over my binder.

The car was a 1932 Austin 10 and the remarkable thing was that it had been modified for the old man, who stood less than 5 feet tall. The pedals were extended and, presumably, there had once been a cushion on the driving seat. However, the old boy still had problems seeing over the bonnet and kept hitting things with the front bumper. To prevent further collision, he took the bumpers off. If you have no bumpers, he assumed, you can't bump into anything. Simple. The son of such a man, I guessed, must have inherited a streak of ruthless logic, and if anyone needed a literal mind untroubled by tangential thought, it was the hero who was going to fix my binder. My invitation was accepted and Farmer Jones and his friend, Frank, arrived next morning.

We dragged the tangled machine out to face the field of golden corn. When working properly, it cuts the corn and, by passing it over a series of conveyor belts, gathers it in a bundle. It puts a string round it, ties a knot in it and spits it on to the ground. That is called a sheaf and the mere sight of one can bring a tear to the eye of those who remember old harvests and festivals. But tears, I suspect, may be shed well before any of our corn is gathered. As I write, the machine has spat its first sheaf in forty years and I must sadly inform you that there is no string round it. A sheaf without a string is like a sausage without a skin — unmanageable. Farmer Jones and Frank are standing like two shocked men at the scene of a nasty accident, trying to work out why.

Much mechanical torment lies ahead before all is safely gathered in.

Wailing cries of despair drifted across our farm for most of this week. Bitterness turned to anger, and language sufficiently foul to make a black pig blush was cutting filthily through the otherwise clean summer breezes. We are still in the middle of our harvest, and if you read my report last week you will know that we are trying to harvest our corn with a binder which has not seen the light of day for forty years. At least before its retirement, the machine had clocked up hundreds of hours of experience; I have none.

As you may remember, our principal preoccupation was

the device which wraps the string round the sheaves and ties a knot in it. I summoned a think-tank of local farmers who still have room in their hearts for these machines, and asked for advice. One suggested emery cloth to get the rust off it, another spun a nut a quarter of a turn, yet another thought the problem might be with the string. I went away to read a book.

It did not help. The paragraph began, "In tying a knot, the knotter bills, which are driven by the cam gear, make one revolution, winding two strands of twine around them while the ends are held fast in the retainer. As the sheaf is ejected, the loop around the knotter bills is stripped off over the two ends grasped between the bills . . ." and that was the simple bit. Baffled, I lumbered back to the field.

I was surprised to be met by smiling faces. It seemed there was progress. Sure enough, as the machine trundled forward, sheaves of corn appeared. I thanked my friends by stuffing a couple of pork chops into their hands, and made plans for my harvest. I decided that having only two carthorses and a binder requiring three, I would pull it with my ancient little Fordson tractor. This turned out to be the best decision of my farming career.

Robert, however, may have made the worst decision of his. He offered to drive the tractor, having been seduced by a romantic photograph in his father's study, showing merry Victorian farm-hands slicing their binder through golden corn on a balmy summer's day. In anticipation of an idyll he turned down a better offer of driving an ultra-modern combine-harvester with stereo radio and air-conditioning.

Off we went. We made two glorious sheaves and then there was a sickening crack. The main drive chain had dissolved into a hundred pieces. We patiently jigsawed it back together and cautiously tried again. Then it jammed. Time was ticking by, the sun was getting hotter. Tempers grew shorter.

Then the canvas ripped. Without the canvas belt, the corn cannot move through the machine. We leapt into a car and drove five miles to throw ourselves at the feet of a man who might repair it. We were back on the field within the hour. Off we went. Ten yards and the chain shattered: we fixed it. The new canvas slipped: we tightened it. Cogs worked loose and we meshed them back together. And then all seemed to be going well. We did almost the full width of the field — and the knotter jammed. We unjammed it and drove a further ten yards till the chain snapped again.

Then my temper broke. I beat the binder with the biggest stick I could drag from the hedge. I cursed it, swore at it with sufficient venom to make its rust drop off and its grease curdle. Robert was thinking wistfully of the air-conditioned cab that could so easily have been his. Still the machine would not go. I almost gave in. I could simply ring a man with a combine for hire and he would whiz round our little patch like a motor-mower on a bowling green. I decided to give it one more try, and if that failed I would pick up the phone and within the day this misery would be over. We pieced the chain together, tensioned the string, revved the tractor. And it worked. The disparate mechanisms sang to the same tune and sheaf after sheaf cascaded like a bubbling mountain stream on to the land. Rejoicing, we broke for tea. When we got back, the chain broke, then the knotter jammed.

Eventually we did it, and it has been the longest week of

my life. I might as well have been cutting the Wembley turf with a pair of nail scissors. So worn down are we that we have not the strength left to rejoice at our victory. The only smiles to be seen are on the faces of our carthorses. They know what they have been fortunate enough to miss.

If you are any good at the obscure type of question they dish out on *Round Britain Quiz*, try this one. When does corn become a shoof, and how many shoofs to a stook, and why would you be shocked when you found out?

You will have gathered that our harvest continues. In my naivety, I had thought that once I had persuaded the binder to do sufficient circuits of the field that would be it. But many mysteries have needed to be unravelled before I can fling my straw hat in the air and declare the job done. The growing mountains of corn are nothing compared with the hoard of wisdom, and bunkum, that our harvest has yielded.

Let us start with the "shoof" called a sheaf everywhere but in Suffolk. It would be wrong, for example, to view it merely as an oversized bunch of flowers. If it has been properly made, the base will have a slant to it so that if you try to prop it upright, it will automatically assume a lean. But how do you know which way the slant has been cut? The moment you pick one up it is sure to become a formless tangle of spiky corn. The answer, my old friends tell me, is to look for the knot, for the relationship between the slant and the position of the knot on the band is always the same. Are you still with me?

If so, we can go on to the next stage, where we pick up the shoofs and make them into stooks. This is the picturesque, and itchy, bit. A field littered with shoofs spat out randomly by the binder is transformed into a living Constable harvest scene. In principle, you pick up two shoofs and lean them against each other, pick up another two and lean them against the first two and carry on till you have done half a dozen. The problem arises when you consider the slant on the base of the shoof; for while one school of thought argues that all shoofs should be stooked with the knots facing outwards, others will put forward philosophical arguments in favour of stooking them with knots inwards. It is easy to discuss this at sufficient length to miss the entire harvest.

Then comes the big decision: which way the stooks should lie. Some will insist a north/south configuration is best so as to get the sun on both sides. Others say east/west is more likely to catch a good drying breeze. Incidentally, the erecting of these stooks is called shocking, which answers my original question.

Our barley was quickly ripe, and so we moved hastily on to the next stage, which is carting it home and building a cornstack. If you were shocked by the erecting of the stooks, be prepared to be bowled over by the intricacies of this apparently simple task. It is vital, the old boys warn me, that when building the load on the cart you leave the centre hollow so the shoofs don't fall off when the cart goes over bumps. However, when building the stack it is vital to do the exact opposite and keep the middle higher so that any rain will run down the straws rather than up them.

But we have had happy days pitching sheafs high into the air to be caught by my old friend who builds the load on the wagon. "Done this often?" I cheekily asked. He paused. "I reckon that must be at least sixty harvests I've worked on." He paused again. "And if you don't pitch them shoofs as I want 'em you won't be seeing me on the sixty-first," and he broke into a great roaring laugh. He was happy to be harvesting the old way. An old farmhand who has been forced to lay up his pitchfork is a kind of exile.

"There's a knack to every job, bar basketmaking," he declared. We said nothing, letting it sink in. Eventually, one of us had to ask what he meant. "Well, basketmaking: you can see through that, can't you?" He laughed till he nearly fell off his load.

The days have been long and hot, but in the evening I have sat contented, and watched the dazzling harvest moon ride high in the sky. It has been our first harvest, and we have won it. It has been a thoroughly shocking experience.

Although what appears here every week is the truth and nothing but the truth, I must confess that if I am guilty of any error it is one of omission.

Four months ago I had our eldest carthorse, Punch, put down. I am sorry not to have told you sooner but, like an embarrassed man with his arm in a sling, I was afraid of repeatedly being asked what happened, and how did it feel? Anyway, I'm still not certain how I feel, for many emotions come into play when a carthorse's life has to end, and not all of them square with the unsentimental approach that livestock farmers must adopt.

I cannot claim that he was the greatest Suffolk Punch that ever lived, but he had led a varied life, working on the streets of Birmingham before graduating to pulling a dustcart around Aberdeen. Star, his old mate from his days in the granite city, is still working on our farm.

Punch was a cunning old devil, and his broad nostrils

could sniff work a mile off: he would pull every trick in the book to put off the evil moment. At harnessing time he would flick his head high just at the moment you were about to drop the collar over his neck, and you would have to swing with all your might on his halter to get him to lower his head. Then, when you turned your back to get his bridle, he'd drop his head to the ground, and if he was lucky his collar would slide off and you'd have to start again.

His worst habit was taking unofficial teabreaks. At one stage in his career, I later discovered, he had been in the charge of a horseman who would take any opportunity to avoid work. Being a cunning old boy, instead of stopping his horses at the end of the furrow and rolling himself a fag there, he would wait until he was half-way across the field. If the farmer saw him from a distance, he assumed something had broken and was being repaired; if he had seen his man and horses lounging at the edge of the field he'd have known they were skiving. This explains Punch's infuriating habit of suddenly stopping: I have on occasions expended more energy on driving that horse along one furrow than on most other farm tasks put together.

Nevertheless, when he took ill with "flu" we were heartbroken. It is a sad sight to see a once proud and upstanding carthorse with hardly the strength to stand, and no will to stagger to his manger, particularly when you remember him fighting his way to his food.

With skilful veterinary care and close attention, he recovered. He regained all his irritating ways, but the sparkle had gone out of him. He fell prey to successive illnesses and in the end was unfit for any work. We kept him for a year before ringing the knacker.

I have spent some months wondering whether it was the right thing to do. I could just as easily have turned him on to the meadow for the rest of his days. Only last night did my conscience clear. I bought Punch from a Suffolk farmer, Roger Clark. He, too, farms with Suffolk Punches and is the kind of instinctive horseman who, although only in his forties, has already forgotten more about it than I will ever know. When he sold Punch and Star to me he said: "They're good horses, but not so good that you won't learn anything from them."

I now know what he meant. An amiable, willing, placid carthorse could almost be worked by a child. You could farm with it for a lifetime and end up no more of a horseman than when you started. But with a tricky horse like Punch you need to be master, and yet sensitive enough to know when he is being idle and when you might be working him too hard. Punch taught me a lot.

And Mr Clark put my mind at rest. "Horses like routine," he said. "Take a farm horse that's been fed three times a day, by the clock, for as long as he can remember. And then when his working life is done, you turn him out in his old age to fend for himself in the field. Alone. That to me is just cruel."

You might disagree but I don't. It would have been pitiful to see the old boy looking over the meadow gate enviously watching the young horses being brought into the yard to be fed and harnessed. I feel I did the right thing by him.

And that's the truth.

CHAPTER NINE

Sex and Seeds and High Society

Autumn 1991

Now that the chill autumn winds are dispersing the last few days of the old farming year, I had planned to take a stroll round the farm and review our position. However, with limbs shaking and vision slightly blurred, I am not certain I could get as far as the farmyard gate. I have succumbed to barley wine. It is home-made and came in an old whisky bottle, thrust into my hand in exchange for a pound of our sausages.

The wine could be easily mistaken for embrocation, or that vile fluid you pour down sheep's throats when they have eaten too much grass and gassed themselves up. I asked my wife to describe the colour of it, my own senses being rather dulled. I hoped she might say it had the subtle hue of a swaying field of ripe barley, but she glumly described it as orange. As to the smell, she declared it to be "Christmas pudding with overtones of compost".

Still, it does help when it comes to putting a rosy glow on what has been a tumultuous and often exasperating year's farming. Since we are in that cosy pause in the traditional farming year, between the harvesting of last year's crops and the ploughing in preparation for next year's, it is a good moment to reflect. (Excuse me while I lubricate the memory with the tiniest drop of the barley wine. Mmm! It does seem to get smoother with every glassful. After a pint, one might even be able to swallow without it hurting.)

Barley, of course, reminds me of the misery of harvesting the cursed stuff. No one ever told me that barley does not grow much higher than daffodils, at least ours didn't. But never mind the short barley, it has been a record year for tall stories. I remember back in the spring, when we were driving an elderly hoe through that barley crop, an old chap who was helping remembered a farm on which he worked as a boy sixty years ago. "Cugh, that ol' farmer, he were an ol' misery, hell of an ol' misery. He used to have a spy hole in the barn and watched us boys working, the old sod did." I have been listening to farmhands' tales for long enough to know that they cannot be hurried.

He turned the horses at the end of the field and settled into hoeing the next row. The story continued.

"Cugh, well one day that ol' booger were walking by the hoss pond an' t' old devil had a stroke. Collapsed he did! Right there by the pond. Course, we didn't go an' 'elp 'im. Cugh! Not after t' way he treated us."

As a son of the new caring society, I was shocked. "Didn't you tell anybody?" I asked in disbelief.

"Well," my friend reluctantly admitted, "we did tell 'is wife but she didn't seem in much of a 'urry to 'elp him either." We laughed. You have to.

The greatest thrill of the farming year has been to see

cornstacks come to life as the horses hauled sheaf after sheaf from field to farmyard. I cannot say they represent the zenith of the art, for I built them myself and a stack-builder's apprenticeship takes decades. "Think of a loaf of bread," I was advised. "Slightly narrower at the base, filling outwards to the eaves, and then a nicely shaped top to let the rain run off." Easy to say, impossible to achieve. It is like trying to build Lego castles with bricks of jelly.

Never mind, another drop of the barley wine can soon correct any deficiencies. In fact, as I look around and see sheep safely grazing, cows placidly awaiting the birth of their first calves, mangel-wurzels swelling into succulent roots, haystacks standing in readiness for the winter, all traumas of the fading farming year seem to disperse.

As my shaking hand raises a glass of barley wine to wish you a Happy New Year, I have made my first resolution: if barley can produce optimistic elixir like this, we must grow even more of it next year.

If world affairs continue down the furrow they are presently ploughing, it may be that our small farm is the last place in the western world where democracy holds no sway. I have been playing dictator again; making declarations and generally putting my foot down.

I have been choosing from our herd of young sows the one that shall be heir apparent to our much-loved Alice. Not that Alice's position as matriarch is in any doubt, but I have to make plans for the awful day when she is no longer able to deliver hardy, thriving litters. So I have been casting an eye over Alice's daughters; and fine upstanding gals they are, too.

I can say that with confidence, for I have before me a list of good points which a young sow should exhibit. Like a judge in the Miss World contest, I took my recipe for the perfect body to the orchard where the four maidens were roaming.

I started as the National Pig Breeders' Association's list suggests, with the head. It should be "well proportioned, broad and clean between the ears" (not as in the real Miss World, where blankness between the ears is an essential). Legs should be "well set, straight and fine-boned". Shoulders should be "fine and in line with ribs". The list warns against "excessive jowl or undershot lower jaw" — presumably to keep out the in-bred county element. Then we come to general movement which should be "active" and — now I'm blushing — the loin which should be "broad and strong" and the belly "full with never less than twelve sound, well-placed teats". While I was trying to take this in, the hungry young sows danced around the trough like flighty models on a catwalk. In the end, I did it the way they do in all beauty contests and went for the one with big features. More than twelve of them.

But it is one thing to nominate a future leader and another to accomplish the *coup d'état*. We now had to separate the chosen one from the rest of the herd who are bound, I'm afraid, for the butcher. I marshalled my forces. As a second lieutenant I needed a tactician who had the measure of the enemy, so I sent for my old friend Dilly, a farmhand of fifty years' experience and a great comfort when moving stock, since there is no disaster which has not befallen him in that half-century.

We set up the gates and hurdles in the corner of the orchard, built a pen, and started the long process of temptation which we hoped would corral three of the four pigs.

We poured some swill and hoped they would be lured into our trap. They were, until they realised what had happened and, with the forcefulness of the SAS storming an embassy, exploded from their temporary prison. We picked ourselves up and started again. Unfortunately, in true Miss World fashion, our contest winner clearly wanted to "travel". Whenever her three doomed girlfriends ambled into the pen, she insisted on joining them. We explained to her that there was literally no future in such behaviour, but she took no notice. We had to pick apples from the trees for her before she was persuaded that out was better than in. As for the other three, greed got the better of them and, together with a combination of several buckets of swill and old Dilly doing an impersonation of a pig that was so realistic I nearly put him in the pen as well, all three were safely despatched.

Of course, Alice knew nothing of all this commotion. She seems to be enjoying life now she is rid of her greedy, brawling offspring. I don't want to unsettle her but, the way things are going in the world, it seems inadvisable to take too long a holiday when others are plotting behind your back. She had better get back to the boar, and get productive again.

The weatherman told us to expect cold showers. Hurrah! I can think of nothing that this farm needs more than an icy downpour. Not only to revitalise the parched earth (as I write we are having our fifty-fourth consecutive day without rain) but to dampen down the lust. We have enough love affairs going on on this farm to supply Barbara Cartland well into the next century, and it is proving a problem. If I were to phrase it in terms an agony aunt would understand, I would now be drafting a letter beginning "I'm worried about my man's performance . . ." and ending with the worrying postscript, "Also, my best friend is making unwelcome advances." If not a bouquet of barbed wire, I fear a posy of electric fencing is developing.

It started last week when we raddled the ram at the start of the tupping season — the time when rams and ewes meet to produce early spring lambs. Despite its suggestive title, this is not a painful fertility rite but simply the process of attaching to the ram's chest a flat wax crayon so that when he covers a ewe, he leaves a lurid greasemark along her back. There is no point in either of them being bashful the morning after, for it is plain to see where the wicked ram has had his way and with how many. As for the ewe, she bears her daub through the winter; and since we change the colour of the crayon every fortnight, you can tell which ewes will lamb first. By the end of tupping you have a weary but contented ram and a technicolour flock of ewes. A lady visitor asked me if they made raddles in a size that would fit her wandering husband. I suggested her crumbling marriage would be better served by a candle-lit dinner, and sold her some pork chops.

Of course, rams, like bulls and boars, can be tricky animals to handle and with this very much in mind we approached our newly delivered boy with care. But he was a pushover. Not only did he stand still while we wrapped the raddle harness round his chest, he even pathetically lifted his legs so we could slip them into the shoulder strap. Once secure, he stood proud and poised like a medalled soldier about to do his duty, or an adolescent lad wearing a dangerous dab of after-shave and off for a good night out.

Amiable though he is, I cannot say he had a devastating effect on the girls. It was days before the tell-tale crayon left its indelible mark. Perhaps he is more the strong and silent type who woos them with his intellect; on the other hand I noticed he chatted them up by sticking his nose up their rear end and then butting them hard if they didn't fall for him straight away. Baa . . . men! The only one he shows consistent interest in is my wife, who now refuses to go anywhere near the meadow.

Worrying about his performance would have been enough were it not for Alice, the Large Black sow, deciding to join in as well. We weaned her most recent litter a few days ago, and would normally have sent her straight to the boar — it is usual practice, and she has never complained. Quite the reverse: she now associates boarding the trailer with making whoopee, and having made that connection we have no trouble getting her into it. However, this time there was no boar for her to go to so Alice has been left fallow.

That was fine for the first couple of days. On day three Alice came on heat. It is a fact that pigs come into season a few days after weaning but we had never experienced this, having left the boar to enjoy the spectacle.

Well, I have never seen anything like it. The sight of old Alice frisking, and barking, and skipping, and sniffing the air for a boyfriend was like seeing your maiden aunt leaning against a bus-shelter and whistling at the sailors.

Feeding became a tricky business. Having decided there was no boar in the vicinity she assumed I must be the next best thing. You may take it from me that being wooed by several hundredweight of black pig is more unnerving than a wink from Mae West. Ignoring the food bucket, Alice rubbed her head violently against my trouser leg, opening her mouth wide and squealing coyly. I ran. She ran after me. She had the look of a woman intent on getting her man. I threw the bucket at her and had the fence not come between us, what followed may well have made anthropological history. The sooner the stock are all under the cold shower, the better.

This is the epic tale of a field of grass: a story of the eternal battle between the young upstart and the elderly but wise remnants of a previous generation.

The story begins last spring, when I decided that we needed a meadow conveniently close to the farmyard. The little three-acre piece was ideal. With the horses we ploughed it, harrowed it and prepared to sow it with grass seed. But (this is where the plot thickens), which seed were we to sow?

The farmer in me wanted a high productivity sward, drought-resistant and fit to stand the wear and tear inflicted by a succession of cows, sheep and heavy-footed horses. The romantic in me had other ideas.

I had been sent a seed catalogue which had on its front cover drifts of daisies amid a profusion of red, blue and golden wild flowers. I turned the page to be confronted with full-frontal moon daisies, corn marigolds and poppies. To a voyeur like me who gets his thrills by peeping through hedges at other people's ancient meadows, this was seriously erotic material. Imagine the excitement then when I discovered that for £200 an acre such a meadow could be mine. I was promised a seed mixture of buttercups, meadow saxifrage, ragged robin and native grasses with names that had my romantic heart pounding: crested dog's tail, meadow foxtail, sweet-scented vernal.

Panting at the thought of it, I was on the point of ordering when my hard-bitten, unsentimental farming streak broke surface and I consulted another catalogue. This had on the front cover fat profitable cattle grazing an impeccable green sward the texture of a carpet. No crimson or purple splashes to spoil it. Just succulent, money-making greenness. The recipes were dreary. Instead of tempting with a whiff of sweet vernal, this catalogue slapped on my plate only "Tetraploid perennial rye-grass, intermediate perennial, and tetraploid late". I could not imagine an old farmworker staggering home with his scythe from the hayfield and sinking into his chair to declare that "Bah, that intermediate tetraploid! That were a sight t' b'hold." Mind you, it's only twenty-five quid an acre, and that is quite a difference.

To cut a long story short I bought a bit of each and did half the field with one and half with the other. I have now been observing the results for the best part of a year.

As you might expect, the young and vigorous modern mixture was up and out of the ground as if it had been fired by some Middle-Eastern supergun. Even in the dry spell it sprouted lush and green while the old, contemplative meadow-mixture wearily raised its head and offered only just enough green to satisfy me that it hadn't passed away peacefully in the night. As the season wore on, the young growth became even more vigorous while the meadow grass slumbered under an ever-thickening blanket of weeds. It became a bit of a local joke. "Do you know, he paid two hundred pound for a bag o' weeds. Bag o' weeds!" The chortle could be heard in the next county.

I watched the stock like a hawk to see which they preferred but I honestly couldn't say they thought better of one than the other. To my eye, however, the two halves of the field are worlds apart. The young upstart is now into his teenage years and has filled out so that not a square inch of bare earth can be seen. The old meadow did have a mad fling in the spring and produce a couple of ox-eye daisies, but that is as far as it wished to go in proving its worth.

Reluctantly I began to suspect that sentimentality has little place in such crucial matters as the choice of grass seed. Old hay meadows must surely be bygones of an inefficient age; no more than weed-studded reminders of the days before the tetraploids landed.

I was about to admit all this to myself when a farming journal proclaimed the results of a study at a Scottish agricultural college: "All Year Round Benefits from Grazing Flower Meadows", it headlined. "Pastures rich in flowers and herbs are providing year-round grazing . . . the information gained could be of vital importance." It was a surprising twist to a tale which I thought had run its course. So I shall watch the tetraploids and meadow grasses once more round the track, like the tortoise and the hare. I hope there will be a moral in the fable, somewhere.

If I made a list of reasons why I should not take a day off from this farm, it would fill an entire book. But to hell with it. In the high autumn, I had a beguiling invitation from a certain "Ploughing and Hedgecutting Society" so I turned my back on my duty, got the lad in to feed the stock, and headed for the Derbyshire Dales.

I have long felt a need for a campaign, modelled on the

one that gave us back Real Ale, in support of the Real Agricultural Show. It may be that I have been unlucky, but the ones I have been to lately are not, strictly speaking, farming events. Having paid heavily to go in and queued for several hours to park the car, you are confronted by a circus of double-glazing exhibits, sullen girls selling striped bodywarmers, enough waxed jacket stands to feed a candle factory and men selling baked potatoes which smell uncannily like the waxed jackets.

Then, if you fight your way past the smarmy estate agents' caravans and crunch over the litter at the entrance to the squalid beer tent, you arrive at row upon row of gigantic farm machinery. It is so far beyond the comprehension, or pocket, of most that it is hardly worth a glance. If you persevere you get to the livestock, about the only bit of farming in the whole show.

Where do I go to find a real agricultural show, a celebration of the countryman's skills? Where can a man stand next to a bale of hay and swell his chest with pride, or rest on his stick beside his mammoth mangel-wurzels and defy any passer-by to say they are not the best ever grown? The answer lies in the Dales.

You know that you are in for a different sort of experience the moment you park the car. You are not confronted by an arrogant youth in an oversized helmet, but by a kindly chap who must be a shepherd, for he has a firm but polite way of waving his crook to move the flock of cars towards vacant spaces. The programmes are 50p, for which he apologises.

The first of the stalls sells nuts and grains to feed calves and sheep; a man sells tractor tyres next to a supplier of galvanised sheepdips. No candyfloss, no beefburgers: just the gentle murmur of farmers screwing down the price of what they want to buy.

The first of the tents houses silver cups, the next has the tweed-suited committee sipping tea, and the third appears to be some kind of shrine. I poked my head in and, sensing no rebuff, I stepped in. Men stood silently, looking towards a stand set like an altar. On it stood sticks with carved wooden handles like crooks, ends made of ram's horn or antler; some were carved, others as they had been cut from the hedge and varnished. We stood in silent awe at the stick-dresser's craftsmanship.

Next came the Ladies Tent, I think. My courage failed me at the last moment, for though this was a seductive marquee humming with chatter and awash with the odours of baking and newly potted jams, I saw no other man in there, and fearing a rebuff from the ladies of the Dales, I retired. Pity. I would have liked to see the entries in Class 2: Three Brown Eggs (Not Broken), or Class 9: A Pot of Beetroot Chutney.

In the arena, shire horses and Clydesdales were ploughing, beyond the hedge, vintage tractors were spluttering as they turned the land two furrows at a time. Nearby, huge modern machines flipped over eight or nine. Above the din could just be heard the sounds of chipping as men with billhooks and saws took straggling lengths of hedge and, using the crafts of the hedge-layer, knitted them into impregnable barriers.

I have never enjoyed an agricultural event as much as this one. It was not about making money, it was a celebration of the skills of countrymen and women. I felt moved to get down on my knees and give thanks, but others were doing it for me. One by one, they shuffled towards the entries in the

Best Bale of Hay contest and knelt as if in prayer. In fact, they were sniffing, and judging, the smells of meadows captured in a way of which bath-salt manufacturers have every right to be jealous.

If you want to know where this gem of an event takes place, I'm not telling. Those cursed double-glazing pantechnicons have ears.

Tucked away in a dusty corner of this newspaper last week was a report of a crime committed in North Yorkshire. Apparently, in the dead of night, thieves armed with forks raided a field of potatoes and made off with the entire crop.

May I make a plea? If anyone knows the identity of these villains, or can give any information as to where they might be hiding, will they kindly give them my address? If it would help, I shall leave forks and spades by the roadside, and some sacks too. I'll even muzzle the dog.

You will gather that we are in the midst of the potato harvest, and the mangel-wurzels are developing that ready-to-be-picked look too. Farming seems to go in phases. I remember the early part of the year when the stock were in the yards and having feed brought to them, I spent months on end with a pitchfork over the shoulder till my neck muscles developed a groove.

Then, in the spring, came the rolling phase when I did nothing but walk up and down the land with the horses, rolling clods; then it was hoeing. Now we're into the backbreaking phase of plucking roots from the ground. To find the entire crop wafted away in the night would be cause for celebration.

I should not really complain: last weekend several hundred people turned up and picked most of my potatoes for me and paid for the privilege. Like all my good ideas, it was someone else's. I remember him saying that if you got a good site near a road and invited people to come and pick their own as they followed the horses along the furrow, you'd draw a crowd.

It worked. I stuck up a few posters, promised potatoes fertilised with only farmyard manure, and elbowed my way on to our local radio station to plug our slogan, "Follow the horses along the furrow and hope to God what you're picking up is a potato!" Eat your heart out, Saatchi's. Sunday morning dawned bright, and with the kettle boiling for 10p cups of tea and the cheese ready sliced for the ploughman's lunches, I waited nervously. By way of insurance I hired some pickers so that if no one turned up, at least the day wouldn't be wasted.

I needn't have bothered. Even before the horses were harnessed a queue was forming. Children, youths, grand-mothers and gentlemen of the cloth stood poised with an array of plastic bags, buckets, and a look in their eye that I have only seen in the few minutes before the start of the Harrods sale.

I called to the horses and our vintage potato spinner inched forward. It is an elderly brute of a machine that cuts beneath the ridges of earth and then, with a series of rotating fingers, spits the spuds out on to the ground.

We had gone only a few yards before the crowd descended on the furrow like gulls following the plough. At the ends of the ridges where the soil is shallower, the potatoes were smaller and there were long faces, but as we worked our way into the crop, giant footballs flung through the air as the spinner moved onwards.

Mothers dumped their babies and scrabbled among the clods of earth like Scarlett O'Hara. The professional pickers watched in disbelief. I developed a barrow-boy's patter. "Come on, ladies and gentlemen, you won't find a fresher spud than these 'ere. C'mon, ten pence a pound."

The day was a great success. Customers were delighted, horses flattered by relentless patting and cooing. The field is bare except for six rows. And now I must face those last few rows alone with no willing helpmates, giggling children or determined old ladies. If the felons wish to pay me a call, we have several left-over ploughman's lunches to offer them.

A finger of conscience that stretches back four centuries has been prodding me. In the writings of my hero, the sixteenth-century farmer Thomas Tusser, I am reminded of my duty at this time of year.

In harvest time, harvest folk, servants and all
Should make all together good cheere in the hall . . .
For all this good feasting, yet art thou not loose,
Till ploughman thou givest his harvest-home goose.

I am clearly required to celebrate that all is safely gathered in. It's not, of course. We have weeks of work ahead of us, but to use that as an excuse would no doubt invoke a sixteenth-century frown. The goose was duly ordered.

Tradition demands it should roam the stubbles and glean the last few ears of corn the harvesters missed. Only when it has eaten its fill is the goose brought to the table. I did consider it, but thought we'd never catch the bird if we let it loose, so we settled for one on the hook. I wanted the harvest table to groan with as much local produce as I could muster: the potatoes came from our own fields, the apples from our orchard (after an argy-bargy with the pigs who live there and think the apple trees belong to them).

My platoon of harvest helpers assembled at seven and, feeling uneasy, we stood around swilling beer and cider. Strange to see these men, whom I had only ever seen before in working woollens and stout farming trousers, smothered in pressed suits. We needed to break the ice. I turned the conversation to mangel-wurzels. An old boy cried out: "Hell, on our farm we only ever used t' grow four mangels." We fell silent. "Yup, just four," he went on. "One in each corner of t' field, and we rolled the beggars home!" Laughs broke out like pigs from a pen, and we sat down to supper.

I once read a book on modern table manners which suggested topics of conversation with which to engage your neighbour. One was: "What novels have you read lately?" Well, I can offer you another: "What's the heaviest sack you've ever carried?" I learnt a lot about sacks from that humble beginning. "Linseed," said Clive, "sacks of linseed are the worst. It's what we call a loving sack, wraps round you, no matter how full you stuff it. I like a good firm sack to get hold of." All nodded and swirled their goose in red cabbage juice.

Then the atmosphere was nearly shattered by a pair of pliers. After supper, I suggested a game in which people brought out small rusty farming bygones and we guessed what they were; a sort of "What's My Line?" You have a device which squeezed, gripped and pulled, and was revealed after some ribaldry to be a 1950s machine for putting rings on rubber teats on a milking machine. After a stud from a carthorse's shoe came the pliers. The jaws were curved, and one had a stub like a small hammer.

"Oh, we had one of those," chimed in Gordon, "we used that to pull the black teeth from piglets. You know, those little sharp ones that jab the sow when they are sucking." You could tell he was in no doubt. Alas, the chap who brought it thought differently.

"No," he said, "that is a bootmaker's tool for turning the leather and hammering home the pin!"

"No, no, no," came the reply, "that's for pigs' teeth!"

Who was right? Had this chap, for sixty years, been using a cobbler's tool for porcine dentistry? Or had a bootmaker been making insensitive use of a delicate veterinary instrument? The argument could have got heated, had not another voice piped up: "On our farm, we only used to grow four mangels . . ." We groaned, and laughed. What a swell party it was.

I decided to avoid all politics after attending a parish council meeting in a village near which we lived. The issue was a well-sprung mattress which had been found on the common just behind the bus-stop. An elderly spinster of many years' council standing thought it was providing marvellous exercise for the youngsters who were too big for the swings; anxious parents of teenagers feared what that exercise might involve. Timidly, I put up my hand and suggested we took it away and burned it. They thought about it. The clerk said we might be in trouble over ownership, someone muttered about our "insurance position". And so it went on, until I decided the chip shop held more attractions than this now wide-ranging debate.

It is only some years later that I feel able to lift my head above the political fence, and once again I expect it to be shot off. Never mind, it is in a good cause.

I was shocked to read last week that European farmers had grown 30 million surplus tonnes of grain: "the largest amount of uneaten food ever produced". It reminded me of an Irish radio report I heard this summer which announced that the entire butter production of that whole country in June was surplus to requirements. It didn't figure very highly in the news because the situation had been the same in May. Add to that the million tonnes of beef in storage (some said to be seven years old) and you have a hell of a beefburger.

Now chew this over: if the ridiculous sums of money being spent on propping up this obscene system were diverted into organic agriculture, could anybody possibly lose by it? Already I hear toast and tea spluttering from the mouths of insulted farmers. I can tell you what they are saying because I have heard it before: "Organic! Can you tell me the difference between the fertiliser that comes out of the back of a cow and the stuff I put on out the bag. Go on!" If you patiently explain that organic growing takes a wider view and encompasses animal welfare, the health of the soil and the sustainability of a farming system which does not depend on the fertiliser lorry coming down the lane, they counter with, "Well, I know a bloke who tried it and got in a right mess." Enquire further and you discover that this bloke's idea of organic farming was simply to plant his seed and do nothing: no understanding of the need to build fertility, no grasp of natural fertilisers. He just switched off the life-support system and expected his crop to live. These are hard men to convert.

But convert they must. It seems to me as I plod along the furrow this autumn that organic farming holds many answers to the problems gripping us by the throat. For a start, it tends to produce a less heavy crop. I hope no one will dare to suggest that that is a bad idea. I hear the mouthful of tea and toast spluttering something about the starving millions around the world, but his system has done little to feed them. It only fills the bellies of redundant aircraft hangars used to store his over-production. To help the hungry we should feed them this year and teach them to farm organically next. And stop encouraging them to grow cash crops for export instead of food for their own nationals.

Farming organically, we would grow less but in a sustainable way. We would fertilise the land by returning to it animal manures (which cease to become the pollutant they are now) and end up with a system which can be sustained without the background throb of an overused chemical industry. It would create jobs in the countryside, too, for organic farming can be more labour-intensive. Animals would have better lives; customers would be delighted.

The spluttering, I guess, has now come round to money; and I don't criticise my imaginary farmer for that. Many are genuinely hard-pressed. But I have already said that the money used to support the present system should be diverted to bolster an organic regime. If it paid the same, who would fail to convert? Only the idle, for organic farming is harder work.

Perhaps there is a moral to be drawn from the fate of the parish mattress. It lay behind that bus-shed for months, its cover rotted to reveal a core of rusty springs. But no one official ever acted. One day a conservation group came to tidy the common and summarily flung it on to a skip. You can do a lot in a little time, if you just stop spluttering about it.

A cow died, and I caught a cold. Two colds really: one adding up to about £600 (the value of the cow) and the other an infection I would have paid a similar sum to be rid of.

This was one of those colds that envelops the skull so that when you bend forward the brain collides with the forehead: no fun when you are picking the last of the potatoes and the final few rows of mangel-wurzels. Since the death of the cow brought on a little head-banging as well, you will gather it has been a gloomy week.

I am presented with a big dilemma. A proper mixed farm, run organically, needs cattle like fruitcake needs raisins. Something has to eat the luscious, fertilising grass mixtures that we sow. Sheep, though useful, do not provide the same mass of invigorating dung. So cattle we must have; but perhaps not the ones we have at present.

They were among the first animals on this small farm: three heifers of the distinguished Red Poll breed. They came from a large herd, and despite all attempts to offer a hand of friendship, they treated us with disdain.

Having come from a smart, professional set-up they probably thought that what we had to offer was a little beneath them.

We hired a bull who spent Christmas with them and duly filled their stockings. In the spring they went out to graze on the marsh and grew ever fatter as motherhood crept up on them.

In September we brought them home and without fuss, in the early hours of a crisp morning, the first presented us with a heifer calf. At sunrise I discovered it, a trembling, slithery little beast lying beneath its confused mother with its first few breaths turning to steam in the cool morning air.

The whole family came to watch it rise to its fragile legs and take its first vital steps in search of breakfast. At last, I thought, there is pleasure and pride in owning these cows.

Then it all started to go wrong. The next calf was a bull, which was fine, but it became increasingly clear that the third cow was not in calf at all. That expanse of girth which we assumed to be offspring was no more than gluttony. A pity, because those who know tell me she is the best-looking cow we own.

The weather turned colder and we brought them in for the winter. I blame myself for what happened next, but let me say by way of mitigation that, with the possible exception of the prime minister, no one receives more unsolicited advice than a novice farmer. Yet in all the words which have come my way, no one told me that cows need magnesium like cars need petrol. It is easy to give, providing you can spot the cow that is short. I did not. I went away for the day and she was found dead by the lad who came to feed.

This left us with an orphan bull calf, who was already a month old. This, believe me, is no cuddly little creature. It may have appealing silky ears and wide innocent eyes, but try to back it into a corner and get some milk down it and you are up against the explosive aggression of a full rugby scrum.

We built a private pen of iron gates but the calf made short work of them and galloped round the farm for two hours, loose. We christened him Ronnie Biggs.

He is a bit friendlier now and no longer needs a tag-wrestling team in full training to administer the pail of milk, but it has been a struggle. By the way, we think he is blind in one eye.

That leaves the heifer calf and her mother. We have never been certain about her, for she has always seemed to be on some kind of brink: behaving more like a mad aunt than the hip-swinging teenage mum she is. She takes offence terribly easily and has a wide-eyed and faintly threatening stare. She is a tiresome beast, and what is worse, her daughter may have inherited the trait.

The night before last I was in the pen having gone five rounds with Ronnie. He had finally quietened and was enjoying a stroke; the setting sun making even more vivid the stunning red coats of my cows.

I thought to myself as I looked at them: I have an orphaned, one-eyed jail-breaker, an aunt and cousin of dubious sanity and a useless, smug, fat maiden. There may be a Booker novel in such a collection of characters, but I fear there is not going to be any profit. Still, a couple of weeks later, Ronnie, the orphan calf, was doing fine. Too fine, really. After a fortnight of being fed by humans he bonded with us to such an extent that I expected him in the kitchen at any minute.

For the first few feeds it took two strong men to hold him, and one to guide Ronnie's head into the milk bucket and imitate his mother's teat with the finger of one hand in the calf's rubbery mouth.

Soon he romped across his pen when he saw the bucket approaching and generally co-operated, preferring to have his ears tickled while he gulped down the milk. I do not think he is for the bullring.

But while all is well with Ronnie, I cannot forget his dead mother. Even in her defunct state she raises an important issue.

Have you ever considered what happens to dead farm animals? Until last year there was a dignified way of getting rid of them. You rang the knacker, who took away the carcass to process it.

He made a little money from selling the hide or the fleece, and the meat for processing into pet food. The offal he sold to another gruesome but essential figure called the renderer, whose factory turned offal into meat-and-bonemeal for fertiliser and manufactured feedstuffs.

The knacker's job was a grim one; dealing day in, day out with injured or diseased animals and disembowelling them for a slender profit. But the system worked. There was enough profit in his cadaverous business to allow him to pay a farmer for the carcass. Farmers were happy, knackers made a living and dead animals were disposed of in a civilised way. But no longer.

Yesterday, instead of a cheque I got a bill for £25, the cost of carting my dead cow away. I can pay it: many farmers can't.

The loss of a good cow is £700 down the drain and farmers cannot afford to lose another penny. But now the old knackering industry is as dead as the animals it deals with, and after speaking to farmers and knackers it is my belief that animals are suffering, public health is at risk and that the official response to a big problem is shameful.

The change stems from mad cow disease and a belief that it was caused by the use of infected offal in manufactured animal feeds. The use of certain offals was, therefore, quite rightly banned. But this meant that the knacker had no way of selling his offal, so instead of making a small profit he had to pay to get it taken away.

Then, in Bristol, a cat was thought to have caught a similar disease after eating manufactured pet food. Hysteria among pet owners ensued and the knacker could now no longer sell the carcass meat to the petfood man. Making no profit, he therefore had to charge the farmer. Many cannot or will not pay.

The official advice to farmers is to bury the corpses, but across huge areas of animal-rearing land in this country the soil is not deep enough. And even if it is, how do you know that your dead and diseased animal is not lying in the path of someone's water supply? Some farmers are simply dumping dead animals by the roadside for councils to clear away.

Knackers have been pleading for government help, but apparently it is like trying to talk to the dead. Perhaps there are simply not enough photo-opportunities in it.

Sick animals are suffering. Let me tell you the tale of one cow, who suffered a difficult calving. The farmer, already hard-pressed, called the vet, whose visit and drugs cost £50. When the cow did not improve he was faced with a dilemma: if the vet put it down, it would mean another bill; he couldn't easily afford the knacker, either. So he decided to wait for it to die and then bury it.

The cow suffered for two long weeks before the farmer's conscience got the better of him and the knacker was sent for. Knackers are hard men, but this one told me he felt deep pity for that cow.

None of this is pleasant, and I apologise if it has spoilt your Saturday morning. But the grim reaping of the knacker is a vital and humane service. There'll never be a flag day to support them, or a society lunch with Liz Taylor to pay their wages. It will always be easier to campaign for the whale or the eagle than for the civilised dispatch of slowly dying animals.

But if you feel moved to mention the issue to the endless dribble of politicians who will be smarming across our doorsteps as election time approaches, then Ronnie's mother will not have died in vain.

One winter day I went into the privacy of my barn and held hands with a farmer. It was a joyous moment as palm met palm in an intimate, rural grasp. We laughed with relief, joked that our worst fears had not been realised and then, in a moment of frankness, confessed that this was something of which we had both dreamt.

You may read into this what you wish: the facts of the matter are that we have at long last brought my mighty threshing machine to life. It is a leviathan that could take our unsaleable sheaves of corn and convert them into profitable grain and straw.

Our relief lay in the fact that the machine had worked; the dreams had actually been nightmares. He dreamt that the five-ton machine had moved itself in the night; I suffered visions of the apparatus collapsing the moment the power was applied. As for the hand-holding, that must for the moment remain a secret between two happy men.

On this farm it is normally the carthorses at plough which make people gasp at the sight of so much strength and energy packed into such a compact frame, but the sight of the threshing machine at full throttle would eclipse a herd of ploughing carthorses. It hums, it whines, it rattles, it can speak if you can understand its language. It is not a machine to be treated lightly: men have lost limbs and lives in the process of extracting grain for people's daily bread.

The machine is a demanding beast, requiring the constant attention of no fewer than eight men. Two, with forks, pitch the sheaves upwards; one man catches them and cuts the string that binds them. Another feeds them gently into the bowels of the monster.

Without mercy, it chews them, beats them, flails them into submission and only when it has finished its violent works does it let the residue go free. The straw is carried out of a wide mouth on to an elevator where it is built into a stack by two more men; another chap places sacks at the grain spouts to collect the precious corn. Yet another poor creature, usually the youngest on the farm, has the unpleasant task of bagging chaff. Chaff is the irritating waste that is too lowly even to qualify as a straw; it is dusty and clinging and anyone foolish enough to wear woollen clothing finds himself looking as dusty as a museum exhibit.

If you consider the straw emerging from the machine's mouth and the grain out of its ears, you will appreciate the miserable orifice out of which the poor boy has to catch the chaff.

I assembled my platoon for nine in the morning and prepared for blast-off. "I reckon it was 'sixty-six we last threshed," said one. "No, I remember it was 'sixty-four 'cos old Isaac had just died and he died in 'sixty-four, because that was the year his sister got caught with . . ." and so the moment of start-up was delayed; nervousness hiding behind banter.

I checked the belt linking tractor to thresher, eased the clutch and slowly as a supertanker she gathered speed. Eyes were glued to each of the fourteen belts and pulleys. I opened the throttle and she groaned at the effort. I revved her until I was certain she sounded steady and then an expert threshing-hand shouted, "Faster!"

I gave her all our little tractor could offer. I felt as if we had the Flying Scotsman in our farmyard, and I was stoking the boiler as the engines surged. She started to shake and roll and hum like a hive of bees. She was ready for us. "Woomph . . ." as the first sheaf went in. She didn't flinch. "Woomph . . ." as the next hit her. Feed too fast and you hear her groan, let her idle and she sighs for more.

We threshed half a stack by lunchtime and drank hot soup at the foot of the growing heap of golden straw. My old farm-worker friend, grinning like a pumpkin at the joy of being back on a threshing gang, shared with us his precious recipe for sugar-beet wine, and memories of its effect.

The grain was carted to the barn in sacks weighing 10 stone or more. Hardly able to handle them, an old farmer grasped me by the hand and showed me how, by making a cradle with your arms, two men can easily carry such a weight between them. We got joy from every sack we dragged from that trailer and even after humping no fewer than fifty, we were still in the grip of it.

You may have noticed that I have been deserting my farm

for one day a week in favour of the television studio. People keep asking me if this heralds my giving up the agricultural life. Nothing could be further from the truth.

When I wake to the prospect of a day's broadcasting, I confess it has its attractions. Instead of having to trudge to the yard and feel the frosty sting of the muck-fork handle, I have the gentle touch of the make-up girl to look forward to. I am thankful that my aged farm-worker friends cannot see me reclining in her chair, each quiff of my hair being shaped into place and shiny skin dabbed dull with a powder puff. "That's what I'd call a rum do!" I can hear them chortling. As for the make-up girls, they invariably ask which Caribbean island gave me my tan. I have tried to explain that following horses along the furrow for windy days on end has the same browning effect as the tropical sun but they suspect I am pulling their leg. So now I say Barbados, and it is no lie. I have renamed one of our fields Barbados to curtail this tedious weekly dialogue.

On studio days, I usually get through to lunch before thoughts of the farm start to sprout and it is often an innocent remark that sets them off. Last week it was the producer who said, "We'll fade to black here." Black! I panic. Black pigs. Alice! How is she? Did the lad turn up to feed her? But on the whole I look forward to a day in front of the lights and return refreshed. But last week things became confused and my hitherto separate worlds collided.

The programme I present is about food, drink and cooking and sensing an opportunity to display his front-man as a concerned food-producer, the editor asked me to bring five legs of lamb which had been reared on this farm so they could be shown and discussed on the programme. I have to confess that lamb has not been selling too well lately so I was happy for an opportunity to get rid of some; and knowing that a mere mention on a television programme has propelled the sales of certain products to stratospheric levels, I seized this as an unashamed marketing opportunity and wondered how many check-outs I ought to plan for in the hut in which we keep the freezer.

I humped the lamb on to the train and penned it firmly in the luggage rack. The thought that they might need some hay for the journey flashed across my hazy mind till I sharply reminded myself that they were long dead and frozen. Yet I was unnerved, sitting in the dreary confines of a railway carriage with the lamb that I had chased, cursed and tended on open sunny meadows throughout spring and summer. Confused thoughts tormented me. As I walked across Liverpool Street station I glanced to check the best places to put up hurdles if the lambs should escape. As we descended the escalator I decided this was the steepest hill they'd slithered down since dipping, which they'd hated. We squeezed on to the Underground and terrified a girl in a short skirt whose bare leg nestled close to the frozen joints. If these lambs had been alive, I pondered, I could have been prosecuted for packing them into transport as crowded as this. We flashed through Green Park and cheered ourselves with thoughts of luscious grazing.

But the worst was yet to come for as the Underground became less crowded I was able to get a seat and, momentarily glancing upwards, the words "Alice, the Large Black Pig" appeared before my eyes. Was I the victim of a ghost, like Marley's? Would I be tormented with visions of farming days past, present and future? The rattle of the train over the points suddenly evoked the jingle of harness, or perhaps the clanking chains of a spirit.

When I had regained my composure, I was relieved to discover on further reading that extracts from this weekly column are being used in an advertisement to promote this newspaper. They are using a bit where I describe the entertaining events surrounding the birth of Alice's most recent litter. I caught a woman reading it, and grinning. I felt an urge to leap across and open wide the carrier bag and show her the legs of lamb, boasting that they were friends of the black pig that was giving her such amusement. On second thoughts it smacked of indecent exposure.

I was relieved to arrive at the studio, hand over the lamb, and put the nightmare behind me. I wandered to the make-up room. "Oh, you look a little pale this morning," said the girl.

"Baa-bados," I replied.

CHAPTER TEN

Cribs, Cattle

Winter 1991–2

Quite accidentally, I have created myself a perfect setting for a nativity play, and it is all due to my hatred of corrugated tin.

Tin roofing keeps out the rain and is cheap, but far from cheerful. It is depressing stuff: it turns barns into slums, stockyards into scrapyards.

The man who invented corrugated tin must have been warped.

But what is the alternative? I have a long, low shed in which I plan to house the early-lambing ewes to protect them from the icy February rains.

I thought of tiles, but they are too heavy. The answer was obvious — I must thatch it. I have home-grown rye straw that spewed from the threshing machine and with a little instruction from my friend Derek, I thought that thatcherism might yet find a place on this farm.

Of course, to some farmers straw is as odious a commodity as tin sheeting is to me. For years it has been common practice to burn the stuff after harvest merely to be rid of it; but on the scale on which I farm, I have always found joy in it.

Red cows amid a sea of newly spread golden oat-straw, seen in a mellow autumn dusk, can bring tears to the eye. Even forking it from stack to wagon to yard has the same sensual pleasure as fluffing duck-down pillows. If some temporary marital dispute ever forced me to sleep the night in the strawed barn, I doubt I would find it any punishment.

My thatching lesson was last Sunday. "First you have to draw y' yealms," said Derek, bending down to grab a greedy handful of straw from the very bottom of the stack. I bent down and grabbed a polite little bundle, which was no use at all.

"Grab more, 's much as y'can grab!" And together we grasped the straw like lodgers in a boarding house fighting over the bread.

Having seized a mighty handful, we then pulled as fast as we could and in one movement the tangled lengths of straw were shaken into neat bundles, all the straws lying the same way. Just to check, we laid them at our feet and flicked through them, discarding weeds and any crumpled stems as we went. This was our yealm.

"I worked for an old farmer," Derek remembered. "He'd never keep cattle under a tin ruf. Allus straw. Reck'nd them didn't sweat up so much and he got more for the hides."

We tied our yealms in neat rows along the roof and having done one layer, went back and did three more.

Half-way through I was struck by a sense of having done this before: a distinct memory emerged of mid-December and a boy sitting in his granny's parlour with a tube of glue and a few straws, making a crib. We made one every year: we painted stones on to cardboard, stuck sand on to papier mâché.

But it was not until we made one with a roof of straw that we proclaimed it our best crib ever. Passers-by would call in and ask to see it.

Standing in my farmyard thirty years later, looking at my shed thatched with straw, happy childhood memories took me back to Granny's Christmas sitting-room. While I was childishly daubing glue, Derek was already working horses on the land, thatching stacks of corn, keeping the hides of the farmer's cows sweet.

And so my lambing shed stands already filled with ghosts of Christmas past. As far as the present is concerned, I can more or less muster enough crib figures. I might just pass as a competent shepherd, although I confess I have done precious little watching of my flock by night.

To be certain of cattle lowing, I merely need to be five minutes late with the cows' evening feed.

We have no kings, but a new young horse called Prince: he will have to do. And, of course, our oldest, most faithful Suffolk Punch must have a part. He is called Star, so I shall stand him in the east.

Which leaves Alice, the Large Black sow, who cannot be left out of this tableau. In the tradition of casting nativity plays out of whatever unpromising material is to hand, and since she has a natural inclination to flop down lazily wherever she finds herself, I fear I may have to put more of my precious straw in the manger and drop her into that. It is the wrapping of her in the swaddling clothes that I am not looking forward to.

While the family were watching the television adaptation of the myth of Theseus and the Minotaur, they were blissfully unaware that their father was pacing the farmyard, fretting like a young Theseus preparing to grapple with the creature that was half man, half bull. My opponent was a fearsome bullock who showed every symptom of having an intelligence greater than mine and who, like the legendary Minotaur, dwelt in a labyrinth few men dared enter. If you are sitting comfortably . . .

Back in the summer when the grass grew high, the clover bloomed, and the mangels swelled to bursting I calculated that we were going to have more winter fodder than mouths to consume it. So I bought a bullock to fatten over the winter. I paid the money and, apart from an occasional visit to his marsh to see that he was well, forgot all about him till last month.

A helpful neighbour cheerily suggested she would pop him into her trailer and bring him home, and I was grateful. But was less cheered when I lifted the phone the following day to be told, "I'm sorry, but we can't catch it. We've been down on that marsh two hours. It's wild. Wild!"

I gathered a platoon of men and we set off to herd the reluctant beast. We might as well have tried to deflect a space rocket as it left the launch pad; for as soon as this demon bullock caught sight of any human, it fled into a labyrinth of fallen trees and scrubland. We went in and flushed it out, edging it nearer to capture only to have it turn at the last moment and charge through our lines as if we were not there.

We held a council of war. Clive suggested the Spanish technique with two wooden balls on a rope which he would swirl around his head and hurl at its rear legs as the galloping beast fled by, to bring it gracefully down. "Have you actually done it before, Clive?" He hadn't. This didn't seem the moment for Clive to relive his Spanish holidays.

Someone thought we ought to lasso it and tie the other end of the rope to the back of a car so it couldn't get away. "And what then?" I asked. No reply. We even debated getting a vet to drug it.

Cheerily, my old farm-worker friend announced, "We 'ad a bullock once. Hell, that was some wild. Three months we tried to catch it. Shot it in the end, with a rifle." We all went home to ponder.

He had to be caught or starve, and I had to convince it that in the cosy farmyard was tasty oat-straw, sugar-beet nuts and juicy mangels. I decided to face it alone.

As I strolled across the marsh in the dusk, the ruddy form was nowhere to be seen. I crept towards his labyrinth making just sufficient noise neither to take him by surprise nor frighten him. I lurched through briars and brambles, but not until I heard his heavy breathing did I discover him lying in a hollow beneath a fallen tree in what was almost a cave. I was carrying a bucket of feed nuts, and looked him in the eye as I held them out. He inched backwards, weighing me up. Then forwards till I was able to snort through my nostrils the way he did through his: I have read this is a good way to make friends. Then I retreated and left the bucket by the gate where I hoped he would come to feed.

It was three days before hunger overcame fear but after a week of these tense encounters, he would trot over whenever I appeared. I was quite fond of him by now.

The phone went. Did I need any help to catch the bullock? "We can get a gang of men . . ." I declined. Theseus went in alone, and so must I. I parked the empty trailer and began the slow process of accustoming him to that.

On the day ordained for his capture, I filled his bucket, put it well inside the trailer, and hid. I watched him inch warily forward, afraid that he would hear the pounding of my heart. One false move, and he would retreat to the labyrinth. I willed his every step forward, prayed for each inch that he moved closer to capture. Then I saw his rump nervously disappear into the box. I crept out of hiding and in the fastest movement I have ever made in my life, slammed the door of the trailer. I could not believe it, and cried out in joy.

He is now safely in the farmyard, and quite happy. But the children are watching the one about the Gorgon whose look could turn men to stone. Our cow, who rather thought she was going to have all the fodder to herself this winter, has been giving me some odd glances.

There will be some new faces round the trough on Christmas Day this year, and a few old friends absent. The farm closes in on itself as midwinter arrives. Cattle, pigs, sheep and horses are in the farmyard now, living as one large, if not entirely happy, family. And whereas on long warm days there were always people dropping in to chat or advise or even help, now that the days are short visitors are rare.

My old farm-worker friend has not been here for weeks now. He has a winter job elsewhere from mid-October until the end of January and is out on the land from eight every morning, when the frost is still hard, until four in the afternoon, when the next frost is starting to form. He picks sprouts: he swipes the top leaves with a broad blade and fells the stalks ready to be carted away.

He does it six days a week: he is sixty-eight years old. I shall think fondly of him as I nudge the Brussels against the turkey leg on Christmas Day.

Alice is not to be found on the farm either. Our Large Black sow has gone to the boar and will not be home until the new year. Alice is a lady of some experience, a mother of thirty-six who knows a thing or two about romance. Her latest boar is a different matter.

I was shocked when I arrived on his home farm to find him a lanky lad of hardly a year old. I am sure he has what it takes, technically speaking, but to see his juvenile frame alongside Alice's mature bulk . . . well, it was like expecting Pinocchio to woo Elizabeth Taylor. I don't suppose he'll know what's hit him. Alice, on the other hand, looks set to have a very merry Christmas, and we look forward to her triumphant return.

The farm goes very quiet just before Christmas. My neighbour keeps turkeys and this is not their lucky time of year. We shall have one on Christmas Day, and I shall be avenged for suffering months of their irritating gabble. Their interminable guttural banter is like living next door to a House of Commons debate, and revenge will be sweet.

But my greatest Christmas wish is that peace may reign in the horse yard. A couple of weeks ago our new horse arrived and we are beginning the steady process of getting him used to his new home and companions. I have made the mistake before of not being careful enough in introducing new horses to the stable, and well remember the dreadful night that Blue arrived and I let him run free in the yard with Star and the now departed Punch.

That no limbs were broken was a miracle. A vicious

fight broke out with three angry horses galloping, bucking, kicking and screaming at each other within the confines of the farmyard. I have never been so terrified.

War raged while each horse fought to establish its supremacy in the new order. This time I took no chances and, although the youngster stands alongside them in the stable, at night he has a separate yard with stout fencing between.

Like the boar, the new horse has much to learn. He has not yet acquired that covering of hearty muscle that will mark him as a true Suffolk Punch, but that will come. And so will a plodding gait: at the moment he tends to trip lightly on his feet and wander as you lead him, like a young child being taken for its first walks.

He has been broken in and is used to the feel of the bridle, the collar and the chains, but it will take experience to turn him into a working horse. A couple of years, at least.

There is no certainty with horses: no guarantee that a young horse will fulfil its early promise or that my plans for him will be fulfilled. I intend to pair him with Blue on the plough when our beloved Star becomes too old for work, but there is no guarantee that they will work as a team.

For the moment they seem happy with each other, standing nose to nose over the fence, nostrils spewing steam in the frosty night air. The older ones are telling him what lies ahead: how long the ploughing days can be, how sweaty the making of hay. But so far, they are not fighting. Conferring, rather, by winter starlight, like three wise men in the East.

New Year, 1992

If you are expecting some cosy rural new year resolutions along the lines of being kinder to butterflies, more sympathetic to hedgerows and speaking softly to the sow, you are going to be disappointed. Agriculture is in the deepest recession, even my little bit of it, and so I face a new year with a harder heart than ever before.

Actually, it might not be all that bad. We had a bumper Christmas with chunky joints of pork being carried off to all parts of the country, and there are only a couple of sacks of potatoes left in the barn. And we have bulging sacks of barley and oats to feed the stock through the coming winter. However, the farm will have nothing further to sell till the spring lambs are fat. So I have made my resolution: I am going to be mean. Not just a bit mean but menacingly mean: Scrooge-like enough to fatten my own granny and send her to market.

The most prosperous farmers have always been the meanest, as far as I can judge. I know of a smallholder who kept a wandering goose that would make occasional marauding visits to a neighbouring field of corn and whose owner was formally warned off by the farmer. Apart from the two square feet of corn which the goose had nibbled, the farmer had a further 700 acres to fall back on. That's what I call mean.

Then there's the one with the hammer. While he was driving home a particularly huge nail, the head parted from the shaft. Rather than go and buy another, he welded the two pieces back together. This would have been laudable were it not that the two halves ended up in a not quite straight line. Imagine the impossibility of hitting a nail on the head when the head of the hammer does not fall precisely where the swing of the hand dictates. But he still won't buy another; for not only would he have to bear the cost of it, he has now spent time and money on welding. As a result, he has got used to it: which means should he ever buy a hammer that hits straight and true, he won't be able to hit anything with it. His brain has been retrained to believe that all hammers are twisted. He's another rich farmer.

But the best example of all comes from the west of Ireland where the peasant farming tradition, which I espouse, still flourishes. It has been said of some of these hard-bitten farmers that they are so mean, if they said the Rosary with you, they'd do you out of a decade. The story goes that a matchmaker struck a romantic deal in which a farmer would agree to marry another farmer's daughter if she came with a dowry of eleven bullocks. The marriage took place but after several years had passed and no children had been born, the matchmaker called at the farm to remind the man that it was his duty to give his wife children and himself an heir. The farmer looked at the matchmaker and said, "Well, you remember her father promised to send eleven bullocks as a dowry the week of the marriage?"

"That's true," said the matchmaker, "'tis all in writing."

"He only sent down nine," said the farmer. "An' until he sends down the other two, I won't lay a finger on the girl." This tale will be my text for the next twelve months. I intend to be like those farmers with whom I seem to be always doing business: the deceptively generous ones who take you into the barn and show you a bit of old machinery, or a heap of useful timber. They ask if you'd like it, and imply that it is a gift. Only when you have expressed interest do they then realise it has some value and the talk eventually drifts round to price.

I am now looking out of the windows, thinking of ways to be mean. I see the carts and think of the money I waste buying grease for the axles. Such profligate spending will have to stop. Goose grease was used for this purpose and so I shall buy a goose (cheaper than a tin of grease from the garage) and fatten it on every bit of waste corn I can find. I shall take it for walks every night along the mangers in the stable so that it can peck up every tiny grain the horses leave behind.

I believe there is a vulgar expression which likens meanness to tightness of a goose's rear end. I shall be studying that rear end, and learning from it.

Even breakfast is under scrutiny. I came into the kitchen after the morning round of stock-feeding thinking about it. In years gone by I might have been a muesli man, but the floury mixture of grains and lumps looks too much like horse-feed to be appetising, and so I converted to toast. But on winter mornings when there might be five hours' heavy ploughing, muck-carting or straw-pitching before lunch, a bellyful of burnt bread cannot fuel this working man. Instead, I have developed a taste for porridge. It makes a mean breakfast. And it occurred to me that it would be an even more mean breakfast if I made it with my own oats.

I was spooning a basinful made from the contents of a bought packet when my gaze fell upon Mr Quaker's informative box. He told me how nutritious it was, how good for my heart, and that it was packed with "soluble fibre". But only when I came to read the list of contents did I realise that here was a perfect opportunity for unparalleled stinginess. A packet of porridge oats, he tells me, contains nothing other than rolled oats. There appears to be no salt, mysterious "stabiliser" or other additives — just rolled oats.

Now, every morning for the past month or so, I have done nothing other than dish out rolled oats to every mouth that would take them. They fuel the horses, vitalise the cattle, and we are now giving them to the sheep to fortify them for early lambing. But in all my humping, scooping and dishing out, never did my mind connect our own oats that came out of the roller with the rolled oats in the porridge packet.

Horrified at missing such a mean opportunity, I went to the barn to roll some for the table. I set the rollers tight so that the husk would be completely crushed, and put the machine in gear. I took a scoop of oats from the bin and fed them gently into the hopper. When it is running empty, the roller squeals as the two halves of the iron mangle rub, but as soon as the oats hit it, it purrs with a crushing hum that persists until the hopper empties. I took my bag of rolled oats back to the kitchen in anticipation of a hearty feast. Sadly, my rolled oats and Mr Quaker's hardly resembled each other. His were white, feathery and floury, and so were mine, but alongside the crushed white kernels lay the bruised and split husks that had been the precious grain's protectors. I had to get rid of the husks before I could call my porridge my own. So near, and yet so far.

I dived into my library of aged farming tomes hoping for a clue as to how this vital separation might be performed. I learnt that porridge bowls should be made of wood, the porridge stirred with a straight wooden stick called a "spirtle", and that to ensure good luck it must always be stirred clockwise. It was the custom to keep the precious oatmeal in a drawer rather than a sack, and farmworkers would ensure it was tightly packed by rolling up their trousers and trampling it with their bare feet.

All this I was prepared to do, but not until I had rid myself of the irritating husks which would make the eating of my bowl of porridge as unpleasant as chewing a dish of empty nutshells.

I tried the winnowing machine. I don't quite know what a winnowing machine is but an old farmer said I ought to have one and he only wanted a fiver for it. (This was before I decided to get mean.) It is the size of a large chest of drawers, has a handle that turns a fan that produces a gale of wind, and sieves and shakes from side to side. This, I thought, might do the trick.

I cranked until the machine shook and its blast of wind whistled like a gale through the barn. I carefully fed the rolled oats into the hopper. But in seconds they were gone. Some flew into the rafters, others settled like newly fallen snow on the hay. Alas, on inspection, I found that although widely travelled, the oat and its husk were still inseparable.

This may seem to be undue effort in pursuit of what many consider a vile, grey and indifferent breakfast fodder. But unless every living soul on this farm gets its oats before the working day begins, a bleak year lies ahead. Once again, I must appeal for your help and advice.

There is one phrase that I have come to loathe. I know it is usually meant kindly, but whenever I hear it I am only too aware of the menace the innocent words conceal. The expression is: "There's a knack to it."

In the traditional type of farming that I am trying to re-create, it is a phrase that occurs all too frequently. In modern farming I would imagine it is hardly ever heard, because high-tech machinery must surely come complete with an instruction manual and no doubt a telephone help-line.

But when you are farming with carthorses and resurrecting machinery that may not have touched the land for fifty years, "knack" appears to be a commodity which is forever in short supply.

I could list 100 jobs that I might accomplish in half the time, if I only had the knack. There is a way of carrying heavy sacks on your back and making light work of it, if you have the knack: if you don't you can displace every muscle from the waist upwards. Every job involving horses is so drenched in knack that it is a wonder it doesn't pour out of the animal's ears. Yet it was a custom among horsemen that they kept their secrets to themselves as it was the only way they could maintain a closed shop: for if they alone possessed the knack, then their jobs were safe.

The result is that students like me have had to struggle to master their secretive art. It is a custom in Suffolk, and probably in other parts of rural Britain, for the wise to keep their silence — but only till it is too late.

I remember in my early ploughing days how I struggled for hour upon bad-tempered hour to adjust my plough to cut a reasonable furrow. It was only as I was walking the horses back to the stable that an old farmhand crept out of the hedge and said: "I've been a-watchin' y'. Trouble is, you 'adn't started right!" This was the sort of creature who would cheerfully watch a man fall from a cliff and then tell him to be careful. I meet a lot like him.

I am struggling at the moment to master a particular knack, and very useful it could be.

I have discovered that before ropes were made out of sisal and Manila hemp, it was the custom of farmers to make their ropes out of their own straw. My 1884 edition of *Stephen's Book of the Farm* gives precise details and boasts "allowing for interruptions, a straw rope of thirty feet can be made in five minutes".

This is indeed good news to a farmer like me who not only has more straw than he knows what to do with, but also resolved that the theme of the current farming year was going to be meanness. Why waste precious cash on coils of unpleasant plastic-covered cordage when in a mere five minutes I could have 30 feet of the stuff for free? All I needed was the knack.

I also required a device known in Suffolk as a "scud-winder" (known in other parts as a "throw-crook" or a "whimble"). I happened to buy one in an auction at the time of the Gulf war and it caused great merriment among the bidders, most of whom thought a scud-winder was going to be some gigantic key that would work a clockwork missile. But in fact it fits neatly in the hand, has a hook at one end and is shaped like a carpenter's brace and bit, so that the hook can be spun with one hand while the fixed end can be held in the other.

The idea is that the hook is engaged in the straw and the twisting begins. As you twist, you edge backwards and a helper then feeds more straw from the stack. I did not have great hopes of success and so didn't bother to summon help. I guessed that if the first part of the operation looked promising, the rest would follow. So my first attempt was alone.

Alone I stood by the straw stack with my scud-winder in my hand and thrust it into the straw till there appeared to be long pieces caught in the hook. I turned the winder and saw the straw begin to spin. It looked promising. Then I edged backwards as instructed and hoped that, like the snake from the charmer's basket, a rope would begin to appear.

It didn't. I ended up with a dangling ball of limp straw that revolved half-heartedly on the end of my whimble like a damp Catherine wheel. I threw it to the ground and plunged the winder once more into the stack, hoping this time I would be blessed with the knack. I spun it and shuffled backwards, but with no success. I spent an hour and gave up. Disheartened once again by lack of rural talent I retired, knackered.

Ever since the suggestion was made that Europe should launch a spy satellite to snoop on farmers and record the movements of livestock, there has been an outcry. Mr Gummer is against it, the Farmers' Union detests the idea, the press think it is a huge joke.

Well, I am all in favour of it; and the closer the scrutiny to which we are subjected the better. It may be considered folly to spend £93 billion to prevent a £27 billion fraud, which is the real reason this satellite is being launched, but for my convenience I think it will be money well spent.

I know it will pose certain problems that will need careful handling: for example, it is the habit of Alice, our Large

Black sow, to relieve herself in exactly the same spot in the corner of the run adjoining her sty.

As she ambles on these discreet little journeys to her corner, she may be far from happy at the thought of being watched. She has her pride. I shall assure her that Brussels officials are well-mannered enough to avert their eyes from their screens when they realise why she is on the move, and that she is not nipping off to collect some fraudulent subsidy.

The real reason I am in favour of it is because it will finally solve what is becoming one of the great farm mysteries. It boils down to this: unbeknown to me, at some stage during the second week of September our ram went missing. I don't know where he went, how he got out, or even how he got back in again. But there is an ever-growing mountain of evidence that this old soldier went absent without leave. For all I know, we may be looking at a case of extreme cowardice.

Way back in August I was being extremely careful about the farm's family planning. I wanted lambing to start in the second week in January. I have a useful little tome called *The Agricultural Notebook*, by Primrose McConnell.

Mine is the 1924 edition and, in passing, mentions an aspect of sheep husbandry called "smearing", which is thankfully extinct. It involved the shepherd working a mixture of Archangel tar and butter into the back of the sheep, having removed the wool first. As you will know by now, I am not entirely in favour of some aspects of agricultural advance, but this is one practice that I am more than happy to let slide into history.

But on the page opposite in the book is a useful chart which tells you when to introduce the ram to the ewes to get lambs at the time you require. It told me that 20 August was the day, and on exactly that date, in he went.

Come the second week in January, we started to lamb. We had a set of triplets from one ewe, twins from another, and then one ewe lost her newborn lamb and one of the triplets was adopted on to her to save its harassed mother from having to suckle three lambs on two teats. Lambing was happening like clockwork — then the clock stopped. This, of course, is not uncommon. Presumably, within the first couple of days of entering the flock, the ram had given each of the girls in season a wink and a nod and then, I assumed, none of the other ewes came on heat (or "on song" as they delightfully call it in Suffolk). However, a sheep's reproductive cycle is sixteen days, so just over a fortnight after the first chorus of passion, the remaining ewes should be singing their heads off again.

Either they didn't, which is unlikely, or the ram was not there when they did. For after the first flush of lambs was born, nothing happened for more than three weeks. In other words, he missed the concert the second time round. So where was he? Had he gone tone deaf? I was around the farm every day over those few weeks and would have spotted him if he had found a gap in the fence. Perhaps he had a headache, or had been reading one of those raunchy Cumbrian novels and gone in search of some younger ewes. Something had clearly taken his mind off the job.

Of course, if the satellite had been on its nosy little orbit, it would have been able to provide the answer to this great mystery. I could have rung Mr Gummer and he would have a print-out of every tupping, complete with duration. The sooner it is launched the better.

In the meantime the ram might need gingering up for next season. I have consulted *The Agricultural Notebook* concerning "tonics". It recommends mixtures of "butter of antimony, Epsom salts, carbolic acid and tincture of myrrh". If that lot doesn't get him joining in the chorus, I wonder what will?

For an entire afternoon the farm took on the look of a Texan oilfield. Rangy, tanned young men arrived in their man-sized trucks. Within moments their brawny arms had transformed a mass of steel piping into a drilling rig and our nineteenth-century farm was beginning to look like the scene of a twentieth-century oil rush.

Before drilling, the rugged young men adjusted their helmets squarely on their heads: I did not have a helmet so I merely twitched my flat cap. As the drill was lowered into the ground, I prayed we would not strike oil. What use would an oil fortune be to me if I had no drop of water with which to make a celebratory cup of tea?

We were drilling for water, deepening an existing well which for as long as anyone can remember has unfailingly provided this farm with water. But I have noticed that on occasions we were drawing water from it faster than it was able to refill itself, and as this part of the world is heading for its driest winter in recent memory it seemed prudent to dig.

If you are in one of those parts of Britain that has been drenched since last August, you may find it difficult to believe that parts of the same islands are in a state of chronic drought. A month ago, in the middle of winter when you might expect the water reserves to be at their fullest, tenants vacated a cottage near here when their well ran dry. I have seen derelict villages on Greek islands abandoned when the springs failed, but never expected to see it on my own doorstep.

For such a vital operation, there seems to be little folklore connected with the digging of wells. We can read endless chapters on the wheelwright, the blacksmith or shepherd, but rarely a word on the mole-like creature who provided that without which no farm or community could have existed.

In his classic descriptions of pre-war East Anglian farming, Adrian Bell, tired of carting water by horse from a pond, hired a man to dig a well for him. He worked by the light of a candle, one at the foot of the well where he dug and another halfway up to test for "bad air". During the digging he was warned by his doctor that old age was overtaking him and that he must "dig no more wells". Rather than leave a hole unfinished, he had a kitchen chair lowered down so that he could catch his breath before he moled further. Bell records that it is the farmer's privilege to descend the well when the digger has finished his work, but my diggers offered no such invitation and I was thankful.

For three days the fresh water flowed again, until one evening I turned the tap in the stable to mix feed for the pigs and not a drop came. Pressured by having to work at new depths, the ageing pump had taken early retirement. It needed more than a sit on a kitchen chair, it needed an expert. We had no water on the farm.

The experience was sobering. Although *we* can manage with a drop for drinking and the merest drain for washing, it cannot be explained to ewes who are feeding lambs that they must hang on until the pump man comes. So my wife fled into town, bought plastic bottles and cadged water from friends, then while I was snugly tucked up in bed she bravely went round the garden water butts at midnight with a bucket to refill the lavatory cisterns.

I rang another well engineer and expected a further assault with high-tech derricks and drilling apparatus. But when he arrived and immediately asked for a light for his candle, I knew I had found the right man for the job. He tied his spluttering light to a rope and lowered it to the bottom, and only when he had satisfied himself that there was no foul air did he descend with his spanners to attack the pump. He tells me he is busy these days, deepening wells, trying to keep pace with increasing drought. I stayed on the surface, lowering cups of tea in the calf-feeding bucket. He did a well nearby, he said, expecting to go down only ten feet. It was thirty before he hit water.

The sound of the revitalised electric pump was like music chiming through the house. The troughs gurgled again, the taps sang. I know that electricity did not come to this farm until the 1950s and that all the water had to be drawn from the well by hand. There are aspects of twentieth-century farming life that I despise but running water is not one of them.

If there is one sound I never want to hear again on this farm, it is the pathetic midnight scrape as my poor wife stumbles, bucket in hand, searching the butts for a drop of water.

Terrific news! I went out late the other night to the barn to get a bucketful of barley. The air was still, the sky clear and every sound for miles around was quite distinct. As I crept through the doors, I heard a rustling like that of a rat. But as I moved further the rustle became more urgent and out of the shadows came a winged white creature with wide eyes piercing down. He circled once above me, eyeing me with his marbly stare, and flew into the night. We have a barn owl in the barn.

I cannot think of anything which has given me more delight. I take it as a seal of approval that the intelligent owl has found a little corner of the agricultural landscape of which he approves or even a barn he finds comfortable.

They don't build barns with crooked oak beams any more. A modern barn is clad in dreary asbestos on a rusting iron framework. Nothing much there for a clawed foot to hold on to, or even a nice knot-hole in which to insert an inquisitive beak. When the rotten walls of our barn were replaced last year, we debated whether it was worth preserving a circular hole in the apex of the roof, about the size of a dinner plate. After an old boy, who years ago laboured on this farm, insisted "you put that there ol' 'ole back in that barn for them ol' owls", we ordered the builder to carve the orifice. It has paid off handsomely.

Of course, to an owl finding a farm like ours must seem like dropping in on heaven. The owl population, I read, was at its greatest in the days when farmers kept their corn in stacks, and hay in ricks; these provided havens for hordes of mice and rats which in turn gave the owl his daily bread. But when the combine-harvester arrived, which processed the grain in the field rather than in the farmyard, the stacks went, the mice became fewer and the owls dwindled.

Well, I have removed all the mousetraps and the rat poison

and our distinguished visitor can now eat away to his heart's content.

I wonder if anyone of influence is ever going to be as wise as the old owl and admit that traditional farming as practised in the first half of this century, for all its financial faults and labouring hardships, was an inherently healthier way of farming the land.

A student came to work here for a few days and on his return to college told one of his tutors that we were growing a crop of vetches; or tares, as they are known in these parts. The student was sharply reminded that this was an old-fashioned crop of no further use, and so not worth preserving. Is this poor lad receiving a rounded education?

Vetches are green-leaved plants, sweet and luscious to graze. They are equally tasty if made into hay or silage to produce high-protein feed. The roots fix fertilising nitrogen in the soil, and the plant's ability to form a dense mat means that any weeds cheeky enough to rear their heads are smothered at birth. So the vetch is a fertiliser, rich feedstuff and effective weedkiller. Nowadays, all those properties could be supplied by applied chemicals. But if you had the choice, how would you prefer to manage your soil? With the virtue-packed vetch, or the questionable drum of chemical? And would you have the arrogance to declare that such a versatile plant "was of no further use"?

I have to admit, having grown vetches last year and found them so dense that the horse-drawn mower could not move an inch through them — my fault for not having sown rye or oats with them to provide an upright stem for them to cling to — I cursed them, too. But a hired tractor-mower did the job, the horse-drawn tedder made the hay, and, having sweated under a crippling sun to stack it, I did curse the day I ever heard of the vetch. But it is now feeding the sheep, who fight for it like children scrabbling for sweets, and the stack will last right through the winter; and I am thankful for it.

I hope that the misguided lecturer was merely having a bad day; but if past form is anything to go by agriculture is not too careful with its precious past.

Were it not for the far-sighted breeders who kept alive the declining species, we would now have no Large Black pigs. No Alice.

At one stage it seemed as if the future of pigs lay in housing them in indoor intensive units. Outdoor pigs were assumed to be "no longer of value". Now, of course, the outdoor pig is in fashion again. But where would modern breeders have gone to acquire the hardiness the modern outdoor pig requires if all the old bloodlines had been allowed to die away?

And vetches, too, will have their day again. But if by that time they are rare, and those who preached their obsolescence try coming here for my precious seed, don't be surprised if I set the barn owl on them. We both know who our friends are.

Indeed, after nearly two years of peasant farming, all of us are getting to know our allies from our enemies. But, hearteningly, the band of allies is growing. And the enemies, if not on the run, are at least learning to keep out of the way of the dotty cow, the storming hogs, the ram and the stick-waving farmer's wife. We are a farm, now. A small one, but well defined, well fenced, and full of hope. And we have our own owl to prove it.

ISIS
55 St Thomas' Street
Oxford OX1 1JG
(0865) 250333

GENERAL NON-FICTION

Estelle Catlett	**Track Down Your Ancestors**
Eric Delderfield	**Eric Delderfield's Bumper Book of True Animal Stories**
Phil Drábble	**One Man and His Dog**
Caroline Elliot	**The BBC Book of Royal Memories 1947-1990**
Jonathan Goodman	**The Lady Killers**
Joan Grant	**The Owl on the Teapot**
Anita Guyton	**Healthy Houseplants A-Z**
Helene Hanff	**Letters From New York**
Dr Richard Lacey	**Safe Shopping, Safe Cooking, Safe Eating**
Sue Lawley	**Desert Island Discussions**
Doris Lessing	**Particularly Cats and More Cats**
Martin Lloyd-Elliott	**City Ablaze**
Vera Lynn	**We'll Meet Again**
Richard Mabey	**Home Country**
Frank Muir & Denis Norden	**You Have My Word**
Shiva Naipual	**An Unfinished Journey**
Colin Parsons	**Encounters With the Unknown**
John Pilger	**A Secret Country**
R W F Poole	**A Backwoodsman's Year**
Valerie Porter	**Faithful Companions**
Sonia Roberts	**The Right Way to Keep Pet Birds**
Yvonne Roberts	**Animal Heroes**
Anne Scott-James	**Gardening Letters to My Daughter**
Anne Scott-James and Osbert Lancaster	**The Pleasure Garden**
Les Stocker	**The Hedgehog and Friends**

The ISIS Reminiscence Series has been developed with the older reader in mind. Well-loved in their own right, these titles are chosen for their memory-evoking content.

Fred Archer	**A Lad of Evesham Vale**
Fred Archer	**Poachers Pie***
Fred Archer	**The *Countryman Cottage* Life Book**
Fred Archer	**When Village Bells Were Silent**
Eileen Balderson	**Backstairs Life in a Country House**
H.E. Bates	**In The Heart of the Country**
Anna Blair	**Tea at Miss Cranston's**
Derek Brock	**Small Coals and Smoke Rings**
Peter Davies	**A Corner of Paradise**
Alice Thomas Ellis	**A Welsh Childhood**
Ida Gandy	**A Wiltshire Childhood**
Lesley Lewis	**The Private Life of a Country House**
Venetia Murray	**Where Have All the Cowslips Gone?**

*Available in Hardback and Softback.

Bill Naughton	**On the Pig's Back***
Diana Noël	**Five to Seven***
Valery Porter	**Life Behind the Cottage Door**
Walter Rose	**The Village Carpenter***
Jan Struther	**Try Anything Twice**
Alice Taylor	**Quench The Lamp**
Alice Taylor	**To School Through the Fields**
Alice Taylor	**The Village**
Flora Thompson	**A Country Calendar**
Flora Thompson	**Heatherley**
Alison Uttley	**The Country Child**
Alison Uttley	**Country Hoard**
Alison Uttley	**Country Things**
Alison Uttley	**Wild Honey**

* Available in Hardback and Softback.

BIOGRAPHY AND AUTOBIOGRAPHY

Lord Abercromby	**Childhood Memories**
Margery Allingham	**The Oaken Heart**
Hilary Bailey	**Vera Brittain**
Winifred Beechey	**The Reluctant Samaritan**
P. Y. Betts	**People Who Say Goodbye**
Christabel Bielenberg	**The Road Ahead**
Kitty Black	**Upper Circle**
Denis Constanduros	**My Grandfather**
Dalai Lama	**Freedom in Exile**
W. H.Davies	**Young Emma**
Phil Drabble	**A Voice in the Wilderness**
Joyce Fussey	**Calf Love**
Valerie Garner	**Katherine: The Duchess of Kent**
Gillian Gill	**Agatha Christie**
Jon & Rumer Godden	**Two Under the Indian Sun**
William Golding	**The Hot Gates**
Michael Green	**The Boy Who Shot Down an Airship**
Michael Green	**Nobody Hurt in Small Earthquake**
Unity Hall & Ingrid Seward	**Royalty Revealed**
Brian Hoey	**The New Royal Court**
Ilse, Countess von Bredow	**Eels With Dill Sauce**
Clive James	**Falling Towards England**
Clive James	**May Week Was in June**

BIOGRAPHY AND AUTOBIOGRAPHY

Paul James	**Margaret**
Paul James	**Princess Alexandra**
Julia Keay	**The Spy Who Never Was**
Dorothy Brewer Kerr	**The Girls Behind the Guns**
John Kerr	**Queen Victoria's Scottish Diaries**
Margaret Lane	**The Tale of Beatrix Potter**
T. E. Lawrence	**Revolt in the Desert**
Bernard Levin	**The Way We Live Now**
Margaret Lewis	**Ngaio Marsh**
Vera Lynn	**Unsung Heroines**
Jeanine McMullen	**A Small Country Living Goes On**
Gavin Maxwell	**Ring of Bright Water**
Ronnie Knox Mawer	**Tales From a Palm Court**
Peter Medawar	**Memoir of a Thinking Radish**
Jessica Mitford	**Hons and Rebels**
Christopher Nolan	**Under the Eye of the Clock** (A)
Christopher Ralling	**The Kon Tiki Man**
Wng Cdr Paul Richey	**Fighter Pilot**
Martyn Shallcross	**The Private World of Daphne Du Maurier**
Frank and Joan Shaw	**We Remember the Blitz**
Frank and Joan Shaw	**We Remember the Home Guard**
Joyce Storey	**Our Joyce**
Robert Westall	**The Children of the Blitz**
Ben Wicks	**The Day They Took the Children** (A)

(A) Large Print books also available in Audio.

BIOGRAPHY AND AUTOBIOGRAPHY

David Bret	**Maurice Chevalier**
Sven Broman	**Garbo on Garbo**
Peter Brown	**Such Devoted Sisters**
Joe Collins	**A Touch of Collins**
Earl Conrad	**Errol Flynn**
Peter Cushing	**An Autobiography** (A)
Quentin Falk	**Anthony Hopkins**
Clive Fisher	**Noël Coward**
Angela Fox	**Completely Foxed**
Sir John Gielgud	**Backward Glances**
Joyce Grenfell	**The Time of My Life**
Reggie Grenfell & Richard Garnett	**Joyce By Herself and Her Friends** (A)
Rex Harrison	**A Damned Serious Business**
Stafford Hildred and David Britten	**Tom Jones**
Teresa Jennings	**Patricia Hayes**
Shirley MacLaine	**Dance While You Can**
Joan Le Mesurier	**Lady, Don't Fall Backwards**
Maureen Lipman	**Something to Fall Back On**

(A) Large Print books also available in Audio.

BIOGRAPHY AND AUTOBIOGRAPHY

Joanna Lumley	**Stare Back and Smile**
Arthur Marshall	**Follow the Sun**
Alma and Ray Moore	**Tomorrow - Who Knows?**
Michael Munn	**Hollywood Rogues**
Elena Oumano	**Paul Newman**
Adua Pavarotti	**Pavarotti**
Harry Secombe	**Arias and Raspberries**
Gus Smith	**Richard Harris**
Terry-Thomas	**Terry-Thomas Tells Tales**
Hilton Tims	**Once a Wicked Lady**
Peter Underwood	**Death in Hollywood**
Alexander Walker	**Elizabeth**
Alexander Walker	**Peter Sellars**
Aissa Wayne	**John Wayne, My Father**
Jane Ellen Wayne	**Ava's Men**
Jane Ellen Wayne	**The Life and Loves of Grace Kelly**
Jane Ellen Wayne	**Marilyn's Men**
Ernie Wise	**Still On My Way to Hollywood** (A)